# Lecture Notes in Mathematics

Edited by A. Dold and B. Eckmann

Subseries: Department of Mathematics
University of Maryland, College Park
Adviser: R. Lipsman

779

# Euclidean Harmonic Analysis

Proceedings of Seminars Held at the University of Maryland, 1979



Edited by J. J. Benedetto

Springer-Verlag
Berlin Heidelberg New York 1980

**Editor**

John J. Benedetto
Department of Mathematics
University of Maryland
College Park, 20742
USA

AMS Subject Classifications (1980): 31Bxx, 42-06, 42A12, 42A18, 42A40, 43-06, 43A45, 44A25, 46E35, 82A25

ISBN 3-540-09748-1 Springer-Verlag Berlin Heidelberg New York
ISBN 0-387-09748-1 Springer-Verlag New York Heidelberg Berlin

Library of Congress Cataloging in Publication Data
Main entry under title:
Euclidean harmonic analysis.
(Lecture notes in mathematics; 779)
Bibliography: p.
Includes index.
1. Harmonic analysis--Addresses, essays, lectures. I. Benedetto, John. II. Series:
Lecture notes in mathematics (Berlin); 779.
QA3.L28 no. 779 [QA403] 510s [515'.2433] 80-11359
ISBN 0-387-09748-1



Printing and binding: Beltz Offsetdruck, Hemsbach/Bergstr.
2141/3140-543210

## TABLE OF CONTENTS

# INTRODUCTION

During the spring semester of 1979 we presented a program in Euclidean harmonic analysis at the University of Maryland. The six lecture series comprising this volume were a major part of our program.

Euclidean harmonic analysis has a rich basic theory and maintains a vital relationship with several other areas which, in fact, have molded the subject and enlivened it with significant applications for over 150 years. Wiener's Tauberian theorem provides a neat example of this fundamental and, to some extent, mysterious interplay. Wiener's theorem not only characterizes the prime number theorem but is used to define spectra properly for phenomena such as white light; this spectral theory provides perspective for the Fourier analysis associated with correlation functions in filtering and prediction problems, and these problems, in turn, lead naturally to $H^p$ spaces.

In the first lecture series of this volume L. CARLESON addressed the two main problems of classical statistical mechanics: a. the verification of expected equilibrium thermodynamic properties and b. the validity of the Gibbs theory for dynamical systems. The results of part a include proofs of the basic properties of the free energy function, as well as a rigorous verification of the existence of phase transition for certain classical models. In part b Carleson first discusses a Boltzmann equation and the approach to equilibrium that it describes. He then considers dynamical properties of harmonic oscillator systems and shows how one can verify the Gibbs theory for an ensemble of such systems. Classical harmonic analysis is pervasive in his approach; and the point of his lectures is to introduce some analytic results and problems which may eventually lead to further progress in applications.

The remaining lecture series contained in this volume, as well as the lectures by our other visitors, fell into one or the other of two categories of problems.

The first category of problems deals with the synthesis of prescribed harmonics to describe a given phenomenon.

The fundamental synthesis problem is to determine whether or not the Fourier series of a function f converges in some designated way to the function. The most famous question in this area treats the case in which f is an element of $L^2[0,2\pi)$ and convergence is pointwise almost everywhere. Carleson answered this question in 1966 by proving that every such function is the pointwise almost everywhere sum of its Fourier series. C. FEFFERMAN gave a conceptually different proof of

Carleson's theorem in 1973, and an explanation of this proof as well as a comparison between it and Carleson's was the subject of his lecture series. Since Fefferman's paper has already appeared (Ann. Math., 98 (1973) 551-571) we have not included his lectures in this volume, and because of this omission we mention a few of his comments. We begin by recalling that in 1968 Hunt proved Carleson's theorem for $L^p[0,2\pi)$, $p > 1$, and that Carleson's method of proof can even be used for the space $L \log L(\log\log L)$. On the other hand, Fefferman's method is $L^2$ in nature and depends on an orthogonality property of linear operators first formulated by Cotlar. To begin with, Carleson's theorem is an easy consequence of the maximal function estimate,

$$(1) \qquad \forall f \in L^2[0,2\pi), \quad \|\sup_N |S_N f(\cdot)|\|_1 \leq C\|f\|_2,$$

where $S_N f$ is the $N^{th}$ partial sum of the Fourier series of $f$. The classical formula, $S_N f = D_N * f$, where $D_N$ is the Dirichlet kernel, expresses $S_N f$ as a Hilbert transform $H$, and the fundamental nature of the operator $H$ in Euclidean harmonic analysis, including its boundedness on $L^2$, provides the basic direction for Fefferman's approach. Instead of substituting the Hilbert transform representation of $S_N f$ into (1), he begins by noting that (1) is equivalent to the estimate, $\|S_{N(\cdot)} f(\cdot)\|_1 \leq C\|f\|_2$, where $N$ is a function depending on $f$ and $x$. Then he observes that (1) follows from the inequalities, $\|T_N f\|_1 \leq C\|f\|_2$, for arbitrary functions $N(x)$, where $T_N f(x)$ is essentially $H(e^{iN(x)y} f(y))$. For each $N(x)$, he verifies the corresponding inequality by making a proper dyadic decomposition of $T$ and applying Cotlar's result to the relatively independent and orthogonal pieces of the decomposition. In his lectures, Fefferman illustrated the method for the case of $N(x) = \lambda x$, which in fact contains the germ of the whole argument; and then, for arbitrary $N(x)$, he explained his combinatorial procedure and decomposition of $T$ into sums of local operators which contain both space and frequency data on small intervals. Regardless of the simplicity or complexity of $f$ or $N$, Carleson's method analyzes the given function $f$ and is oblivious to the corresponding function $N$, and Fefferman's method does the opposite.

Synthesis was also the subject matter of the lecture series by both Y. DOMAR and L. HEDBERG. The problems they discussed fall into the category of spectral synthesis and have the following formulation: let $X$ be a class of distributions with support contained in a fixed subset $E$ of $\mathbb{R}^n$; determine whether or not a given element $\mu \in X$

is the limit in some designated topology of bounded measures contained in X. In Domar's case the Fourier transform of X is a subset of $L^{\infty}(\mathbb{R}^n)$ and the topology is weak * convergence. This is the setting of Beurling's classical spectral synthesis problem based ultimately on Wiener's Tauberian theorem. Domar considers the case in which E is a curve in $\mathbb{R}^2$ and he characterizes spectral synthesis in terms of the curvature of E. He also solves some analogous problems for manifolds E in $\mathbb{R}^n$, $n \geq 3$, and obtains spectral synthesis results in terms of the geometric properties of E. In Hedberg's case, X can be any one of a large collection of Sobolev spaces and the topology is the Sobolev space norm topology. This is the setting in which the spectral synthesis property for all elements of X is equivalent to the stability, in the sense of potential theory, of closed sets essentially complementary to E. Hedberg verifies this equivalence in various Sobolev spaces, and analyzes and generalizes Wiener's criterion for regular points in order to characterize Sobolev space spectral synthesis.

The second category of problems deals with the harmonic analysis of operators of $L^p$ spaces. These problems have emerged from the research of Zygmund, Calderón, and Stein, as well as several of our guests. The omnipresent Hilbert transform H and its generalizations are an essential feature of the area, and multipliers, maximal functions, $H^p$ theory, and interpolation are some of its major topics.

In order to verify various $L^p$ estimates for H and related operators, R. COIFMAN and Y. MEYER presented a range of real and complex methods, from Boole's symbolic calculus of over a century ago to the latest $\bar{\partial}$ proofs of Calderón's theorem. Boole's theory systematically uses measure preserving maps and has long been a staple for ergodic theorists; in harmonic analysis it provides a means to calculate the distribution function of H. Large parts of Coifman's and Meyer's lectures were given in the context of commutators and bilinear maps. Commutators of H are used in the study of boundary value problems for elliptic equations, and they arise naturally when one wishes to extend the classical $L^2$ estimate for H to curves. Next, G. WEISS, in joint work with several others, set forth a theory of interpolation which includes the Riesz-Thorin theorem and Stein's theorem for analytic families of operators. He dealt with a continuum of Banach spaces associated with the boundary points of a domain $D \subseteq \mathbb{C}^n$ and constructed intermediate spaces for each point of D. The basic interpolation result is stated in terms of subharmonic functions. An interesting corollary of the theory is an extension of the celebrated Wiener-Masani theorem which, in turn, provides important factorization criteria for certain filtering and prediction problems. Finally, A. CÓRDOBA solved several specific problems involving a thorough mix of many of the real methods in this second category of problems and concepts. The first result settles a basic real variable question on the differentiation of integrals and depends on a covering theorem and estimates on the appropriate maximal function. The remaining results include a rather complete theory for multipliers arising from classical summability methods.

We wish to thank Berta Casanova, Cindy Edwards, Pat Pasternack, Becky Schauer, and June Slack, of our technical typing staff for their expert work; and to express our appreciation to Alice Chang, Robert Dorfman, Ward Evans, Raymond Johnson, and C. Robert Warner for their editorial assistance.

John J. Benedetto
College Park, Maryland

Besides many of the analysts at the University of Maryland, the participants in our program included:

| | | |
|---|---|---|
| M. Ash | L. Ehrenpreis | A. Picardello |
| A. Baernstein | E. Fabes | H. Pollard |
| M. Benedicks | C. Fefferman | E. Prestini |
| G. Benke | R. Fefferman | F. Ricci |
| R. Blei | A. Figà-Talamanca | L. Rubel |
| G. Bohnké | L. Hedberg | C. Sadosky |
| P. A. Boo | H. Heinig | D. Sarason |
| L. Carleson | R. Hunt | P. Sarnak |
| P. Casazza | C. Kenig | P. Soardi |
| L. Cattabriga | T. Koonwinder | A. Stray |
| R. Coifman | J. Lewis | J.-O. Stromberg |
| A. Córdoba | L. Lindahl | N. Th. Varopoulos |
| L. de Michele | L. Lipkin | G. Weiss |
| J. Dieudonné | Y. Meyer | G. Woodward |
| Y. Domar | C. Mozzochi | R. Yamaguchi |
| P. Duren | D. Oberlin | M. Zafran |

# SOME ANALYTIC PROBLEMS RELATED TO STATISTICAL MECHANICS

Lennart Carleson
Institut Mittag-Leffler

Apology. In the following lectures, I shall give some analytic results which derive from my interest in statistical mechanics. I do not claim any new results for applications, and any serious student of statistical mechanics should consult other sources. It is my hope that analysts will find, as I have, that interesting and difficult analytic problems are suggested by this material; and that they will eventually make contributions of real significance in applications.

## I. Classical Statistical Mechanics. Background

1. We consider a system of $N$ particles moving according to a Hamiltonian function

$$H(p,q) = H(p_1,\ldots,p_{3N},q_1,\ldots,q_{3N}).$$

The classical equations for the motion are

$$\text{(1)} \qquad \dot{q}_i = \frac{\partial H}{\partial p_i}, \quad \dot{p}_i = -\frac{\partial H}{\partial q_i} .$$

$p_i$ are the momenta and $q_i$ the position coordinates for the particles. It follows that $H(p,q)$ is constant during the motion and $H$ is interpreted as the energy of the system. A typical situation is

$$H(p,q) = \frac{1}{2}\sum p_i^2 + \sum_{i\neq j} \Phi(\tilde{q}_i-\tilde{q}_j), \qquad \tilde{q}_i = (q_{3i+1}, q_{3i+2}, q_{3i+3}).$$

We now assume that the motion takes place inside a box $\Lambda_N$ of volume $\sim\rho^{-1}N$, where $\rho$ is the density of the particles. The total energy $H = E_N \sim \lambda N$, so that $\lambda$ is the average energy per particle.

Denote by $d\sigma$ the surface element of the energy surface $\Sigma_N$ in the 6N-dimensional space $\Omega$ of points $\omega = (p,q)$.

The basic assumption of statistical mechanics is now that the motion $\omega(t)$ is ergodic on the energy-surface, i.e.,

$$\lim_{N\to\infty} \lim_{T\to\infty} \frac{1}{T}\int_0^T \varphi(\omega_N(t))dt = \lim_{N\to\infty} \frac{\int_{\Sigma_N} \varphi(p,q)d\sigma}{\sigma(\Sigma_N)}$$

at least for simple functions $\varphi$ depending on a finite number of variables and belonging to $C_0^\infty$. Actually, from a physical point of view it is more natural to assume that

$$\lim_{T\to\infty} \frac{1}{T}\int_0^T \varphi(\omega(t))dt$$

exists, where $T$ avoids a set of density zero. We then speak of the Gibbs limit. A natural assumption here is also that we are dealing with a bounded number of different particles, and therefore have a corresponding number of symmetries in the function $H$. I shall not formulate this in more detail; the meaning in concrete cases is quite clear.

Gibbs' contribution here is that he has given a formula for computing the density $d\sigma/\sigma(\Sigma) = d\mu$. Let us observe that

$$d\omega = d\sigma\, dE \quad \text{in} \quad \Omega_N.$$

Let $\beta$ be a parameter and consider

$$F(\beta) = \int_\Omega e^{-\beta E}\, d\omega = \int_0^\infty e^{-\beta E}\, dV(E)$$

where $V(E_0)$ is the volume $\int_{E<E_0} d\omega$. By partial integration

$$F(\beta) = \beta \int_0^\infty e^{-\beta E} V(E)\, dE.$$

The dependence on $N$ is now such that

$$E = Ne, \qquad V(E) = v_N(e)^N C_N, \quad \text{where} \quad v_N(e) \longrightarrow v(e),$$

and $v(e)$ is expected to be a smooth function. We are dealing with an integral essentially of the form

$$I_N = C_N \int_0^\infty e^{-N[\beta t - \psi(t)]}\, dt$$

where $\psi(t)$ is an increasing function bounded from above. If we define

(2) $$-\psi^*(\beta) = \sup_t (\psi(t) - \beta t)$$

we realize that

$$I_N(\beta) \leq e^{-N\psi^*(\beta)} \cdot \text{Const.}$$

and

$$I_N(\beta) \geqq C_N \int_{t_0}^\infty e^{N\psi(t_0)} \cdot e^{-N\beta t}\, dt = \text{Const.}\ \frac{e^{-N\psi^*(\beta)}}{N}.$$

Hence $I_N$ and so $F(\beta)$ get their essential contribution from the surface $t$ where the supremum is taken.

$\psi^*(\beta)$ is the Legendre transform of $\psi(t)$. Observe that

$$\psi(t) + \psi^*(\beta) \leqq \beta t.$$

Hence

$$\psi^{**}(t) \geq \psi(t)$$

and $\psi^{**}$ is the smallest convex majorant of $\psi$.

Only those values of $\beta$ which correspond to linear pieces in $\psi^{**}$ give ambiguous values of $t$ in (2). We have

$$-\psi^*(\beta) = \psi(t) - \beta t \quad \text{and} \quad \psi'(t) = \beta$$

so that

$$\psi^{*\prime}(\beta) = t \quad \text{if} \quad \psi'' \neq 0.$$

If the graph of $\psi^{**}$ contains a straight line then $\psi^*$ shows a corner. Hence, if $\psi^*$ is smooth, then $\psi^{**}$ is strictly convex.

Going back to $F_N(\beta)$, the proper definition is

$$f(\beta) = \lim_{N\to\infty} \frac{\log F_N(\beta) - \log C_N}{N} .$$

Unless the energy surface is one of the exceptional values for which we have ambiguity in $t$ we can choose $\beta$ so that the integral in the definition of $F(\beta)$ is carried out essentially on the right energy surface. If

$$\mu_{\beta,N} = \frac{e^{-\beta E} d\omega_N}{F_N(\beta)}$$

then it follows that

$$\int \varphi(p,q)\,d\mu = \lim_{N\to\infty} \int \varphi(p,q)\,d\mu_{\beta,N} ,$$

and this is Gibbs' rule. We also see that we can expect exceptional results if $f(\beta)$ has a singularity—in these cases it is not clear that the formula gives the correct result.

In the case of the simple Hamiltonian,

$$\frac{1}{2} \sum p_i^2 + \sum \Phi(\tilde{q}_i - \tilde{q}_j),$$

the first integral over $p$ gives

$$C^N \beta^{-\frac{3}{2}N} .$$

Classical thermodynamics tells us that we should interpret $\beta$ as an inverse temperature. The second part is

$$\int_{\Lambda_N} \!\!\cdots\! \int e^{-\beta \Sigma \Phi(\tilde{q}_i - \tilde{q}_j)} dq_1 \ldots dq_{3N} .$$

It was proved only rather recently by Ruelle that $f(\beta)$ does indeed exist in a case like this. The problem of regularity of $f$ is, however, still unsolved. Here we shall give some related results, noting that $f(\beta)$ is always analytic for small $\beta$.

The problems we have dealt with are closely related to a problem in probability, viz., the problem of large deviations. This was studied by Cramér and Feller and the following results which we need later are well known.

Let $X_1, X_2, \ldots, X_N$ be real stochastic variables with identical distribution and assume

$$E(e^{\lambda X}) = F(\lambda) < \infty .$$

We are interested in

$$\mathrm{Prob}\left(\sum_1^N X_j > tN\right) = e^{-N\mu_N(t)}, \qquad t > E(X).$$

Clearly,

$$F(\lambda)^N = -\int_{-\infty}^{\infty} e^{\lambda t N} d\left(e^{-N\mu_N(t)}\right) = \lambda N \int_{-\infty}^{\infty} e^{N(\lambda t - \mu_N(t))} dt .$$

Hence,

$$e^{N \log F(\lambda)} \geq \lambda N \exp\{N(\inf_t(\lambda t - \mu_N(t))\}\int_{-\infty}^{0} e^{N\lambda t} dt \simeq e^{N\mu_N^*(\lambda)}$$

and

$$e^{N \log F(\lambda)} \leq e^{N\mu_N^*(\lambda)} \int_0^{N^2} dt = N^2 e^{N\mu_N^*(\lambda)} .$$

Therefore,

$$\mu_N^*(\lambda) = \log F(\lambda) + 0\left(\frac{\log N}{N}\right).$$

Since $\log F(\lambda)$ is smooth it follows that

$$\lim_{N\to\infty} \mu_N(t) = \sup_\lambda(\lambda t - \log F(\lambda)).$$

In a similar way one can compute high moments

$$E\left(\left(\frac{X_1 + \cdots + X_N}{N}\right)^{aN}\right) \sim e^{bN}, \qquad E(X) > 0.$$

One finds that

$$b = a \log a - a \log \lambda - a + \log E(e^{\lambda X})$$

$$\lambda \frac{E(Xe^{\lambda X})}{E(e^{\lambda X})} = a .$$

2. In the general case, the motion described by (1) is extremely complicated. Boltzmann introduced a random element in the description of the motion, which is highly plausible but just as difficult to verify. The classical theory concerns elastic collisions between particles assumed to occur in a random fashion. Here we shall present an extremely simple but at the same time very general model which contains some of the characteristics of the Boltzmann theory.

Suppose we have a system of $n$ particles, each in one of $N$ states, $N \ll n$. We should think of the state as a given position and velocity. At each time the particles in states $(i,j)$ can interact and go over to the states $(\nu,\mu)$. The proportion of particles $(\nu,\mu)$ which arise in this way is

$$A_{ij}^{\nu\mu} p_i(t) p_j(t) \Delta t,$$

where $p_i(t)$ is the proportion of particles in state $i$ at time $t$. The matrix $A_{ij}^{\nu\mu}$ is assumed to safisfy

(3) $$A_{ij}^{\nu\mu} = A_{ji}^{\mu\nu} = A_{\nu\mu}^{ij} \geqq 0, \qquad (i,j) \neq (\nu,\mu).$$

We set

(4) $$A_{\nu\mu}^{\nu\mu} = -\sum_{(i,j)\neq(\nu,\mu)} A_{ij}^{\nu\mu}.$$

For $p_\nu(t)$ we obtain in this way the differential equations

$$p_\nu'(t) = \sum_{i,j,\mu} A_{ij}^{\nu\mu} p_i(t) p_j(t),$$

which is a general discrete Boltzmann equation. It has many features of the usual equation, and the proofs are, of course, all very easy.

(A) $$\sum_1^N p_\nu(t) \equiv 1.$$

Proof. By (4) it follows that

$$\sum_1^N p_\nu'(t) = \sum_{i,j,\nu,\mu} A_{ij}^{\nu\mu} p_i(t) p_j(t) \equiv 0.$$

(B) $$p_\nu(t) \geqq 0.$$

Proof. Suppose first that $\alpha_\nu = p_\nu(0) > 0$ for all $\nu$ and that if $p_\nu(t) = 0$ then $p_\mu(t) \neq 0$, $\mu \neq \nu$. By analyticity this set of $\alpha_\nu$' s is dense. Suppose now that $p_\nu(t_0) = 0$ and $p_i(t) > 0$ for $0 \leq t < t_0$ and for all $i$. Then

$$p_\nu'(t) = p_\nu(t) \sum_{j,\mu} (A_{\nu j}^{\nu\mu} + A_{j\nu}^{\nu\mu})\, p_j + \sum_{i,j\neq\nu,\mu} A_{ij}^{\nu\mu} p_i p_j,$$

i.e., an equation $p_\nu' = \varphi p_\nu + f$, where $f \geq 0$ on $(0,t_o)$. Hence,

$$g(t) = p_\nu \exp\left\{-\int_0^t \varphi d\tau\right\}$$

is non-decreasing on $(0,t_o)$. Since $g(0) > 0$, it follows that $p_\nu(t_o) > 0$ which is a contradiction. The general case follows from density.

(C) $H(t) \equiv -\sum_1^N p_\nu(t)\log p_\nu(t)$ is non-decreasing.

<u>Proof</u>. 
$$H'(t) = -\sum_{i,j,\nu,\mu} A_{ij}^{\nu\mu} p_i p_j \log p_\nu$$
$$= -\frac{1}{2}\sum A_{ij}^{\nu\mu} p_i p_j (\log p_\nu + \log p_\mu)$$
$$= -\frac{1}{2}\sum A_{ij}^{\nu\mu} p_i p_j (\log p_\nu + \log p_\mu - \log p_i - \log p_j)$$
$$= -\frac{1}{4}\sum A_{ij}^{\nu\mu} (p_i p_j - p_\nu p_\mu)\log \frac{p_\nu p_\mu}{p_i p_j} \geq 0 .$$

There is equality if and only if $p_i p_j = p_\nu p_\mu$ whenever $A_{ij}^{\nu\mu} \neq 0$.

(D) Let $\Lambda$ be the linear space of vectors $\lambda = \{\lambda_\nu\}_1^N$ such that

$$\sum_{\nu=1}^N \lambda_\nu p_\nu(t) = \sum_{\nu=1}^N \lambda_\nu p_\nu(0)$$

for any choice of initial values $p_\nu(0)$. $\Lambda$ is called the invariants of the motion. In classical theory they are the moments and the energy. Here we first have the trivial invariant $\lambda = \{1\}$.

$$\lambda \in \Lambda \text{ if and only if } A_{ij}^{\nu\mu} \neq 0 \Rightarrow \lambda_i + \lambda_j = \lambda_\nu + \lambda_\mu .$$

We can therefore interpret $\Lambda$ as an additive invariant under possible interactions.

<u>Proof</u>. Assume $\Lambda$ satisfies the condition. Then

$$\sum_1^N \lambda_\nu p_\nu'(t) = \sum A_{ij}^{\nu\mu}\lambda_\nu p_i p_j = \frac{1}{2}\sum A_{ij}^{\nu\mu}(\lambda_\nu + \lambda_\mu - \lambda_i - \lambda_j) p_i p_j = 0.$$

Assume, conversely, that $\sum A_{ij}^{\nu\mu}\lambda_\nu p_i p_j \equiv 0$ for all $p_i \geq 0$ for which $\sum_1^N p_i \equiv 1$. We may also assume that $\sum\lambda_\nu = 0$. It follows that the quadratic form has to be a constant multiple of $(\Sigma p_i)^2$, i.e.,

$$\sum_{\nu,\mu} A_{ij}^{\nu\mu}(\lambda_\nu + \lambda_\mu) = C.$$

Consider

$$\sum A_{ij}^{\nu\mu}(\lambda_\nu + \lambda_\mu - \lambda_i - \lambda_j)^2 = \sum A_{ij}^{\nu\mu}[(\lambda_\nu + \lambda_\mu)^2 + (\lambda_i + \lambda_j)^2]$$

$$- 2 \sum A_{ij}^{\nu\mu}(\lambda_i + \lambda_j)(\lambda_\nu + \lambda_\mu).$$

The first sum vanishes. The second equals

$$-C \sum_{i,j} (\lambda_i + \lambda_j) = 0.$$

Hence,

$$\lambda_i + \lambda_j = \lambda_\nu + \lambda_\mu \quad \text{if} \quad A_{ij}^{\nu\mu} \neq 0.$$

(E) Let us now assume that the system is "ergodic" in the following sense. Let $E$ be any set of indices. Let $\overline{E} = \{\nu \mid \exists \mu$ and $i,j \in E$ with $A_{ij}^{\nu\mu} \neq 0\}$. Then the system is called ergodic if for any set $E$, $E_1 = \overline{E}$, $E_2 = \overline{E}_1, \ldots, E_k = \overline{E}_{k-1}$, and $E_k$ = all indices for $k$ large enough. We choose $t_n \to \infty$ so that $p_\nu(t_n) \to \pi_\nu$. By (C),

$$\pi_i \pi_j = \pi_\nu \pi_\mu \quad \text{if} \quad A_{ij}^{\nu\mu} \neq 0. \tag{5}$$

Let $E$ be the set where $\pi_i \neq 0$. If $i,j \in E$ and $A_{ij}^{\nu\mu} \neq 0$ it follows that $\nu,\mu \in \overline{E}$. Hence, $E = \overline{E}$ and it follows that $E$ = all indices, i.e., $\pi_i \neq 0$ for all $i$.

We have

$$-\sum \pi_\nu \log \pi_\nu = H(\infty)$$

and

$$\sum \pi_\nu \lambda_\nu = \sum p_\nu(0)\lambda_\nu.$$

By (5), $\log \pi_\nu$ is an invariant, i.e.,

$$\pi_\nu = \exp\{-\sum_\lambda c(\lambda)\lambda_\nu\}.$$

Finally, let $x_\nu$ solve the extremal problem

$$\sup(-\sum x_\nu \log x_\nu), \quad \sum x_\nu \lambda_\nu = \sum \pi_\nu \lambda_\nu \text{ and } \lambda \in \Lambda.$$

By the Lagrange theory we have

$$x_\nu = \exp\{-\sum_\lambda d(\lambda)\lambda_\nu\},$$

and $x_\nu$ is unique by Jensen's inequality. We have

$$\sum x_\nu \log \pi_\nu = \sum \pi_\nu \log \pi_\nu \quad \text{and} \quad \sum \pi_\nu \log x_\nu = \sum x_\nu \log x_\nu$$

since $\log \pi_\nu$ and $\log x_\nu$ are invariants. Hence,

$$\begin{aligned} 0 &= \sum(\pi_\nu \log x_\nu - x_\nu \log x_\nu + x_\nu \log \pi_\nu - \pi_\nu \log \pi_\nu) \\ &= \sum(x_\nu - \pi_\nu)\log \frac{\pi_\nu}{x_\nu} \le 0, \end{aligned}$$

which gives $\pi_\nu = x_\nu$.

Let us summarize the result in a theorem.

<u>Theorem</u>. Let $(A_{ij}^{\nu\mu})$ be an ergodic transition matrix. The limits,

$$\lim_{t\to\infty} p_\nu(t) = \pi_\nu,$$

exist and $\pi_\nu > 0$. $\{\log \pi_\nu\}$ is an invariant and $\{\pi_\nu\}$ maximizes the entropy $H$ for all distributions with given invariants.

## II. The Harmonic Oscillator

1. We consider a model where a particle $P_\nu$ is placed at each point of a lattice. Many results would be true in the several dimensional case but for simplicity, let us assume that the lattice is $\mathbb{Z}$. The particles make small oscillations and the movement is governed by the Hamiltonian

$$H_N(p,q) = \frac{1}{2}\sum_1^N p_\nu^2 + \sum_1^N a_{\nu-\mu} q_\nu q_\mu = \frac{1}{2}|p|^2 + U(q).$$

We assume $a_\nu = a_{-\nu}$ and that $U(q) \ge 0$, i.e.,

$$A_N(x) = \sum_{-N}^{N} a_\nu e^{i\nu x} \ge 0.$$

When $N \to \infty$, $A_N(x) \to A(x)$, and we assume $a_\nu \to 0$ sufficiently rapidly.

The Gibbs' theory is in this case trivial. The free energy is

$$N \log F_N(\beta) = \log\left\{\int e^{-\frac{\beta}{2}|p|^2} dp \int e^{-\beta U(q)} dq\right\} = -N \log \beta + C_N$$

so that $F(\beta) = C\beta$. The connection between energy and $\beta$ is simple. We write

$$\int e^{-\frac{\beta}{2}|p|^2}dp = c\int_0^\infty r^{N-1}e^{-\frac{\beta}{2}r^2}dr = c\int_0^\infty e^{-\frac{\beta}{2}r^2+(N-1)\log r}dr.$$

The main contribution to the integral comes from

$$r \sim (N/\beta)^{\frac{1}{2}}$$

so that

$$\frac{1}{2}r^2/N = \frac{1}{2\beta},$$

i.e., the kinetic energy/particle is $1/2\beta$. The same computation for the potential energy yields

$$\frac{1}{N}U(q) = \frac{1}{2\beta}.$$

The energy is, therefore, in equilibrium, equally divided between potential and kinetic energy.

To study the time evolutions we have to consider the equations

$$y_\nu''(t) = -2\sum_\mu a_{\nu-\mu}y_\mu(t)$$

writing $y_\nu$ for $q_\nu$. Assume for simplicity $y_\nu(0) = 0$ and set $y_\nu'(0) = b_\nu$. We assume $|b_\nu| \le 1$ or somewhat more generally $|b_\nu| < C|\nu|^c$ and that

$$\sum |a_\nu||\nu|^c < \infty.$$

$y_\nu(t;N)$ denotes the solution and we set

$$M_N(t) = \sup_{s\le t}\sup_\nu \frac{|y_\nu(t)|}{|\nu|^c}.$$

Standard methods give

$$M_N(t) \leqq C\sup\frac{|b_\nu|}{|\nu|^c}e^{Ct}.$$

From this we see that $\lim_{N\to\infty} y_\nu(t;N) = y_\nu(t)$ and that $\mathbf{y(t)} = \{y_\nu(t)\}$ is a unique solution.

The solution $y(t)$ is easily described explicitly. Let $b(x)$ denote the distribution

$$b(x) = \sum_{-\infty}^{\infty} b_\nu e^{i\nu x}$$

and let

$$a(x) = \sqrt{2A(x)}.$$

Assuming first $A(x) \geq \delta > 0$, we see that $Y(x;t) = \sum_{-\infty}^{\infty} y_\nu(t)e^{i\nu x}$ is a distribution and

$$Y(x;t) = b(x) \frac{\sin a(x)t}{a(x)}.$$

This formula makes sense also when $A(x) = 0$, using the power series for the right hand side. Hence

$$(1) \qquad y_\nu(t) = \frac{1}{2\pi} \int_{-\pi}^{\pi} b(x) \frac{\sin a(x)t}{a(x)} e^{-i\nu x} dx,$$

using integral notations for distributions. The corresponding formula for $y'_\nu(t)$ is

$$y'_\nu(t) = \frac{1}{2\pi} \int_{-\pi}^{\pi} b(x)\cos(a(x)t)e^{-i\nu x} dx.$$

As a simple example, take $b_\nu = 2\cos \alpha\nu$, i.e., $b = \delta_\alpha + \delta_{-\alpha}$. Then if $a(\alpha) \neq 0$,

$$y_\nu(t) = b_\nu \frac{\sin a(\alpha)t}{a(\alpha)}, \quad \text{for all } \nu.$$

Hence $y(t)$ stays on a very special trajectory and is not at all distributed as the Gibbs measure. Observe however that

$$\frac{1}{P} \sum_{1}^{P} \frac{1}{T} \int_{0}^{T} y'_\nu(t)^2 dt \to \frac{1}{P} \sum_{1}^{P} \frac{b_\nu^2}{2} \sim 1, \quad P \to \infty,$$

so that the result on kinetic energy holds. To obtain the Gibbs theory we therefore need to assume some symmetry on the initial values but the special result on kinetic energy may hold more generally.

For this reason let $B$ be a random variable with distribution $F(b)$ and assume

$$\int_{-\infty}^{\infty} b dF(b) = 0, \quad \int_{-\infty}^{\infty} b^2 dF(b) = 1.$$

Let $b = \{b_\nu\}_{-\infty}^{\infty}$ be an independent sequence from $B$ and let $y(t;b)$ be the corresponding solution. In accordance with the discussion in the introduction we shall say that Gibbs theory holds if, given any weak $\varepsilon$-neighborhood in the space of measures in $R^{2n+2}$, there is a $T(\varepsilon)$ so that, for any $T > T(\varepsilon)$, the probability (in the initial value distribution) that the distribution of $(y_0, y_0', \ldots y_n, y_n')$ on $(0,T)$ does not fall in this $\varepsilon$-neighborhood of the Gibbs distribution is $<\varepsilon$. What this means computationally is soon clear. The following theorem holds.

Theorem. For each fixed $z$ consider the condition

(2) $\qquad a'(x) = z$ only on a set of measure zero.

If (2) is valid for either of the cases,

(a) for all $z$ if $F \neq \frac{1}{\sqrt{2\pi}} e^{-b^2/2}$, (b) for $z = 0$ if $F = \frac{1}{\sqrt{2\pi}} e^{-b^2/2}$,

then the Gibbs theory holds for the harmonic oscillator A and the initial values B.

Lemma. Assume condition (2) (a) and let $\Lambda$ be a polynomial. Then

$$\sup_{\nu} \left| \int_{-\pi}^{\pi} \Lambda(x) e^{-i\nu x} \cos(a(x)t)dx \right| < \delta$$

for all $t$, $\delta T \leq t \leq T$, if $T > T(\delta)$.

Proof. (Lemma) Subdivide $(0,2\pi)$ into $K$ intervals $I_j$, $K$ fixed, and let $T \to \infty$. By (2), $a(x)$ is bounded below except in $o(K)$ of the intervals $I_j$.

In each interval $I_j$ consider $(\nu,t)$ so that

(3) $$|\nu - a'(x)t| > K^2\sqrt{T}\ , \ t > \delta T,\ x \in I_j.$$

For each choice of $(\nu,t)$ the inequality (3) holds for all but $o(K)$ intervals $I_j$ - uniformly in $(\nu,t)$ - unless $a'(x) = \text{constant} = \alpha$ on a set of positive measure. If (3) holds a partial integration shows that

$$\left| \int_{I_j} e^{-i\nu x} \Lambda(x) \cos(a(x)t)dx \right| \leq \frac{\text{Const.}}{K^2}$$

Proof. (Theorem) We consider the distribution function $H_T$ of $y_0', y_1' \ldots y_p'$ on $(0,T)$ and compute its characteristic function. We restrict ourselves in the first place to the derivatives to simplify the formulas.

Then we have

$$M_T = \int e^{i\sum_0^P \lambda_j \xi_j} dH_T(\xi_0, \ldots \xi_p) = \frac{1}{T} \int_0^T dt \, \exp\left\{ i\sum_0^P \lambda_j \frac{1}{2\pi} \int_{-\pi}^{\pi} b(x)\cos(a(x)t)e^{-ijx}dx \right\}.$$

Letting $\Lambda(x) = \sum_0^P \lambda_j e^{-ijx}$, the expression under the exponential sign can be written

$$\sum_{-\infty}^{\infty} b_\nu \frac{1}{2\pi} \int_{-\pi}^{\pi} \Lambda(x)e^{-i\nu x}\cos(a(x)t)dx$$

$$= \sum_{-\infty}^{\infty} b_\nu \frac{1}{2\pi} \int_{-\pi}^{\pi} (\Lambda_1\cos \nu x + \Lambda_2\sin \nu x)\cos(a(x)t)dx.$$

$M_T$ now is a function of the initial values. If we observe that

$$(4) \qquad \int_{-\infty}^{\infty} e^{ibu} \, dF(b) = e^{-\frac{1}{2}u^2 + o(u^2)} \qquad , u \to 0$$

and also that

$$\sum_{\nu=-\infty}^{\infty} \left( \frac{1}{2\pi} \int_{-\pi}^{\pi} (\Lambda_1\cos \nu x + \Lambda_2\sin \nu x)\cos(a(x)t)dx \right)^2 =$$

$$= \frac{1}{2\pi} \int |\Lambda|^2 \cos^2(a(x)t)dx,$$

we find, using the lemma,

$$(5) \quad \lim_{T\to\infty} E(M_T) = \lim_{T\to\infty} \frac{1}{T} \int_0^T e^{-\frac{1}{4\pi} \int_{-\pi}^{\pi} |\Lambda|^2 \cos^2(a(x)t)dx} \qquad dt.$$

If $F$ is the normal distribution, the lemma is not needed since there is no error term in (4). Now, if $a(x) = c$ only on sets of measure zero, it follows from Wiener's theorem on measures without point masses that

$$\lim_{t\to\infty} \int_{-\pi}^{\pi} |\Lambda|^2 \, 2\cos(a(x)t)dx = 0 ,$$

if we avoid a set of $t$ of density zero. Hence

$$\lim_{T\to\infty} E(M_T) = \lim_{T\to\infty} \frac{1}{T} \int_0^T e^{-\frac{1}{8\pi} \int_{-\pi}^{\pi} |\Lambda|^2 dx} \qquad dt = e^{-\frac{1}{4}\sum_0^P \lambda_j^2}$$

which is the Fourier transform of Gibbs' distribution.

To prove the theorem we must also compute the second moment of $M_T$.

The computation is completely analogous and will not be repeated here. The result is

$$\lim_{T\to\infty} E(|M_T|^2) = \lim_{T\to\infty} (E(M_T))^2.$$

From Tchebycheff's inequality we deduce that, for any given $T$ large enough,

$$\left| M_T - e^{-\frac{1}{4}\sum_0^P \lambda_j^2} \right| < \varepsilon$$

except on a set of initial values of probability $< \varepsilon$. For any finite set of $\lambda$'s the same is true and we have proved convergence in the sense specified.

The correct condition on $a(x)$ is no doubt that $a(x) \neq zx + w$ on sets of measure zero. I have however no proof of the lemma in which this condition is sufficient.

To prove ordinary ergodicity, we need some extra assumptions, partly on $F$ to control the error term in (4) and partly on $a(x)$ to control the lemma uniformly. The following theorem holds.

Theorem. If, in addition to the earlier assumptions,

$$\int_{-\infty}^{\infty} |b|^{2+\delta}\, dF(b) < \infty, \quad \text{some} \quad \delta > 0,$$

and if

$$\dim[a'=0] < 1$$

then the system of derivatives is ergodic.

We now turn to the computation for the distribution of $y(t)$. The computation is the same up to the formula (5). Here we find for the similar characteristic function $M_T^*$ that

$$(6) \qquad \lim_{T\to\infty} E(M_T^*) = \lim_{T\to\infty} \frac{1}{T}\int_0^T e^{-\frac{1}{4\pi}\int_{-\pi}^{\pi} |\Lambda|^2 \frac{\sin^2(a(x)t)}{a(x)^2}\, dx}\, dt.$$

We have to distinguish two cases.

Theorem. The Gibbs theory holds for the complete distribution if and only if - given earlier conditions on $a$ -

$$(7) \qquad \int_{-\pi}^{\pi} \frac{dx}{A(x)} < \infty.$$

One could easily describe the precise situation if (7) does not hold. However, let us be content with a study of the case when

$$A(x) = x^2 \quad \text{at} \quad x = 0$$

and $A(x) \neq 0$ otherwise. Then we should change the scale of the $\lambda$'s and replace $\lambda_j$ by $\frac{\lambda_j}{\sqrt{T}}$. The expression in the exponent of (6) changes to (at $x = 0$)

$$\frac{1}{4\pi}\int_{-\pi}^{\pi} |\Lambda(x)|^2 \frac{\sin^2 xt}{x^2}\frac{dx}{T}$$

and we get convergence to

$$\lim_{T\to\infty} E(M_T^*) = \int_0^1 e^{-\frac{1}{4}\sigma|\Lambda(0)|^2} d\sigma.$$

The particles therefore line up in a row with

$$y_0(t) \stackrel{\sim}{=} y_1(t) \stackrel{\sim}{=} \cdots \stackrel{\sim}{=} y_p(t) \stackrel{\sim}{=} \sqrt{T},$$

and the distribution is a combination of normal distributions. On the scale 1 we have, on the other hand, independence since the derivatives are independent.

In dimension two a similar phenomenon occurs but on scale $\sqrt{\log T}$, while in 3 and more dimensions, the Gibbs theory in general holds.

2. We now turn to the second problem, concerning the distribution of energy. The problem is surprisingly difficult and it is an open question as to what extent the conditions on $a(x)$ and $b$ are necessary.

Theorem. Suppose that $A(x) \in C^5$ and that $A'(x) > 0, 0 < x < \pi$. Suppose that $b$ is a pseudomeasure ($|b_\nu| \leq C$), that $0,\pi \notin \operatorname{supp}(b)$, and $b(x) = b(-x)$. Suppose finally that the correlations

$$\lim_{N\to\infty} \frac{1}{2N+1}\sum_{n=-N}^{N} b_n b_{n+k} = \rho_k, \quad k = 0,1,2\ldots$$

all exist. Then for all $\nu$

$$\lim_{T\to\infty} \frac{1}{T}\int_0^T y_\nu'(t)^2 dt = \frac{1}{2}\rho_0.$$

Proof. It is of course sufficient to consider $\nu = 0$. By (1)

$$y_0'(t)^2 = \frac{1}{4\pi^2}\iint b(x)b(y)\cos(ta(x))\cos(ta(y))dxdy.$$

Let $w(u) \geq 0$ be an element of $C_0^\infty$, assuming $\int_{-\infty}^{\infty} w(u)du = 1$ and $w(u) = w(-u)$. Let $\hat{w}$ be its Fourier transform. Then

$$\frac{1}{T}\int_0^T y_0'(t)^2 w(t/T)dt = \frac{1}{8\pi^2}\iint b(x)b(y)\hat{w}(T(a(x)-a(y)))dxdy$$
$$+ \frac{1}{8\pi^2}\iint b(x)b(y)\hat{w}(T(a(x)+a(y)))dxdy.$$

Since $a(x) + a(y) \geq \delta > 0$ on the support of $b(x)b(y)$, the last integral is easily proved to tend to zero using localization. We shall now also use localization in the first integral. It is then enough to study $x,y > 0$, by symmetry. Observe first that

$$(8) \qquad a(x) - a(y) = a'(\tfrac{x+y}{2})(x-y) + O((x-y)^3).$$

Let $h_T(u)$ be a function with support in $|u| < 2T^{-1+\delta}$, $h_T(u) \equiv 1$ in $|u| < T^{-1+\delta}$, and $\|\hat{h}\|_1 \leq C$. If $|x-y| > T^{-1+\delta}$ then for any second derivative $D^2$ and $\pi-\delta \geq x,y \geq \delta$

$$|D^2\hat{w}(T(a(x)-a(y)))| < C_N T^{-N} \quad \text{for all } N.$$

Hence

$$\|((1-h_T(x-y))\hat{w}(T(a(x)-a(y))))^\vee\| < CT^{-N}.$$

We may therefore restrict the first integral by multiplication by $h_T(x-y)$. If $|x-y| < T^{-1+\delta}$ we also have

$$|D^2(a(T(a(x)-a(y))) - a'(\tfrac{x+y}{2})(x-y))| < T^{-1+\delta}.$$

We may therefore also replace $a(T(a(x)-a(y)))$ by the similar expression from (8) and we may drop $h_T$ by the same argument as above.

Finally we may introduce a function $\phi(x) \in C_0^\infty$ which has support strictly inside $(0,\pi)$ and is $\equiv 1$ in a neighborhood of the support of $b(x)$. The result is that we should prove convergence as $T \to \infty$ of

$$(9) \qquad \sum b_\nu b_\mu \int_{-\pi}^{\pi}\int_{-\pi}^{\pi} \phi(\tfrac{x+y}{2})h(\tfrac{x-y}{2})\, e^{i\nu x+i\mu y}\, \hat{w}(Ta'(\tfrac{x+y}{2})(x-y))dxdy$$

where $h = h_{T_0}$ for some fixed $T_0$. We introduce the new notation, $x+y = 2\xi$, $x - y = 2\eta$, $\nu + \mu = n$, $\nu - \mu = m$, and have to compute

$$W(m,n;T) = \iint \phi(\xi)h(\eta)e^{in\xi+im\eta}\hat{w}(2Ta'(\xi)\eta)d\xi d\eta.$$

Observe now that $a'(\xi) \geq \delta > 0$ where $\phi \neq 0$. Hence, for all $N > 0$,

$$\frac{1}{2\pi}\int_{-\infty}^{\infty} h(\eta)\hat{w}(2Ta'(\xi)\eta)e^{im\eta}d\eta = w\left(\frac{m}{2a'(\xi)}\right)\frac{1}{2a'(\xi)} + O(T^{-N}).$$

Thus, we obtain

$$(10) \qquad W(m,n;T) = \int_{-\infty}^{\infty} \phi(\xi)e^{in\xi}\, w\left(\frac{m}{2a'(\xi)T}\right)\frac{d\xi}{2a'(\xi)T} + O(T^{-N}).$$

Observe that besides the estimate (10) we have

$$(11) \qquad |W(m,n;T)| \leq \frac{C}{Tn^2} + O(T^{-N}),$$

$$(12) \qquad |W(m,n;T)| \leq C\,\frac{T^4}{(n^2+m^2)^2}.$$

We write (9) as

$$\sum_{m=-\infty}^{\infty}\sum_n b_{\frac{n+m}{2}}\, b_{\frac{n-m}{2}}\, W(n,m;T) = \sum_{|m|<T^4}\sum_{|n|<T^4} + \sum_{(\text{Rest})}.$$

The second sum is easily estimated by (12). In the first sum we replace $W(n,m;T)$ by (10) and can omit the remainder term, leaving us with

$$\sum_{|n|<T^4}\sum_m b_{\frac{n+m}{2}}\, b_{\frac{n-m}{2}}\int_{-\infty}^{\infty}\phi(\xi)e^{in\xi}\, w\left(\frac{m}{2a'(\xi)T}\right)\frac{d\xi}{2a'(\xi)T}.$$

Observe now that the inner sum only extends over $|m| < CT$. We therefore have the trivial majorant

$$\sum_{|n|<T^4}\sum_{|m|<CT}\frac{C}{Tn^2}$$

and can therefore compute the sum termwise as $T \to \infty$ for $n = 0, \pm 1, \pm 2 \ldots$. By assumption, we have

$$\sum_{m<s} b_{\frac{n+m}{2}}\, b_{\frac{n-m}{2}} = \rho_n s + o(s).$$

Hence,

$$\sum_{|m|<CT} b_{\frac{n+m}{2}}\, b_{\frac{n+m}{2}}\int_{-\infty}^{\infty}\phi(\xi)e^{in\xi}\, w\left(\frac{m}{2a'(\xi)T}\right)\frac{d\xi}{2a'(\xi)T}$$

$$= \int_{-\infty}^{\infty}\phi(\xi)e^{in\xi}\left(\rho_n\int_{-\infty}^{\infty} w(u)du + o(1)\right) \to \rho_n\int_{-\infty}^{\infty}\phi(\xi)e^{in\xi}\, d\xi,$$

and our limit becomes

$$S(\phi) = \sum_{-\infty}^{\infty} \rho_n \int_{-\infty}^{\infty} \phi(\xi)e^{in\xi}\, d\xi.$$

To compute this sum, let us take $\psi$ with support in $|\xi| < \delta$ outside the support of b. Then, for k large enough, we have

$$\left| S(\psi) - \frac{1}{k} \sum_{j=1}^{k} \sum_{n} b_{n+j}\, b_j \int_{-\infty}^{\infty} \psi(\xi)e^{in\xi}\, d\xi \right| < \varepsilon.$$

The inner sum can be written

$$\frac{1}{k} \sum_{j=1}^{k} b_j \sum_{n=-\infty}^{\infty} b_n \int_{-\infty}^{\infty} \psi(\xi)e^{-ij\xi}\, e^{in\xi}\, d\xi = 0.$$

Hence $S(\phi)$ is independent of the definition of $\phi$ outside the support of b. We have $\phi \equiv 1$ in a neighborhood of the support. Hence

$$S(\phi) = S(1) = \rho_0.$$

Checking the constants for some trivial choice of b, our result follows.

3. We shall now study in some detail the mechanics of the situation, i.e., assume that $b_\nu \to 0, |\nu| \to \infty$, in some suitable sense. To get the connection with classical analysis let us again assume that $A(x) \sim x^2$ and is otherwise $\neq 0$. If we consider the formula (1) for $y_0(t)$ we see that

$$y_0(t) \sim \frac{1}{2} S_t(0;b) \ , \ t \to \infty,$$

where $S_t$ denotes the $t^{th}$ partial sum of the Fourier series of b. It is of course no coincidence that Gibbs was interested in overshoots in situations of this type.

If however A(x) is bounded below, the situation is radically different. Suppose for simplicity that A(x) has its only extreme points at $0,\pi$. Then, for b(x) smooth, we have

$$y_0(t) \sim C_1 \frac{b(0)}{\sqrt{t}} + C_2 b(\pi) \frac{\sin a(\pi)}{\sqrt{t}} .$$

It is of interest to study asymptotics of this type for as general functions b(x) as possible. Only the local behaviour of a(x) at $x = 0,\pi$ is of importance. Therefore, $x^2$ is completely general for all a(x) with $a''(0) \neq 0$. We get the following problem: Let $f(t) \in L^1(-1,1)$ and study

$$f_N(x) = C\sqrt{N}\int_{-1}^{+1} e^{iN(x-t)^2} f(t)dt, \ N \to \infty.$$

This is also of basic interest for the study of spherical summability of Fourier integrals. In $\mathbb{R}^2$, as an example, the relevant kernel is

$$\sqrt{N}\ \frac{e^{iN|x-t|}}{|x-t|^{3/2}}, \quad x,t \in \mathbb{R}^2.$$

It will turn out that even continuous functions have pathologies, and we shall limit our discussion to Hölder classes $H_\alpha$, $0 < \alpha < 1$. We may assume $f \in H_\alpha$ on $(-\infty,\infty)$ has compact support. Then $\hat{f}(\xi)$ exists,

$$f(x) = \int_{-\infty}^{\infty} \hat{f}(\xi)e^{i\xi x}\, d\xi$$

in a suitable sense, and

(13) $$f_N(x) = \int_{-\infty}^{\infty} \hat{f}(\xi)e^{i\xi x}\, e^{i\xi^2/4N}\, d\xi.$$

Since $\hat{f}(\xi) \in L_1$ if $f \in H_\alpha$, $\alpha > 1/2$, $f_N(x) \to f(x)$ uniformly in this case. Actually, a slightly better result holds.

<u>Theorem</u>. Suppose that $|f(x) - f(y)| = o((x-y)^{1/2})$ uniformly. Then $f_N(x) \to f(x)$ uniformly. If $f \in H_{1/2}$ then $f_N(x)$ is bounded but $f_N(x)$ need not converge.

<u>Proof</u>. Assume first the o-condition. Fix $N$ and define $a_\nu$ by the relations

$$Na_\nu^2 = 2\pi|\nu|, \text{ where } a_\nu < 0 \text{ if } \nu < 0 \text{ and } a_\nu > 0 \text{ if } \nu > 0.$$

Consider $x = 0$ and assume $f(0) = 0$. We have

$$f_N(0) = C\sqrt{N}\sum_{-N}^{N}\left\{\int_{a_\nu}^{a_{\nu+1}} e^{iNt^2}(f(t)-f(a_\nu))dt + \int_{a_\nu}^{a_{\nu+1}} f(a_\nu)e^{iNt^2}dt\right\}.$$

Now observe that

$$\left|\int_{a_\nu}^{a_{\nu+1}} e^{iNt^2}\, dt\right| = O\left(N^{-\frac{1}{2}}(|\nu|+1)^{-\frac{3}{2}}\right).$$

The second sum is dominated by

$$\sqrt{N}\sum O\left(\frac{|\nu|^{1/4}}{|N|^{1/4}}\right)O\left(N^{-1/2}|\nu|^{-3/2}\right) = N^{-1/4}\sum_{1}^{N}\nu^{-5/4} \to 0.$$

In the first term we use the assumption on $f$ and find the estimate

$$\sqrt{N}\sum_{-N}^{N} o(|\Delta a_\nu|^{3/2}) = N^{-1/4}\sum_{1}^{N} o(\nu^{-3/4}) = o(1).$$

If, instead, $f \in H_{1/2}$, we see from the proof that $f_N(x)$ is bounded. To see that $f_N(x)$ need not converge, take a sequence $N_j$ so that $N_{j+1}/N_j \to \infty$ and define

$$f(t) = \sum_{1}^{\infty} N_j^{-1/2} e^{-iN_j t^2}.$$

It is easy to see that $f(t) \in H_{1/2}$. For $f_{N_j}(0)$ we have the expression,

$$f_{N_j}(0) = C\sqrt{N_j}\sum_{k=1}^{j-1} N_k^{-1/2}\int_{-1}^{+1} e^{i(N_j-N_k)t^2}\,dt + 2c + \sum_{k=j+1}^{\infty} O(N_j^{1/2}N_k^{-1/2}),$$

and so

$$\lim_{j\to\infty} f_{N_j}(0) = f(0) + 2c.$$

If we allow an exceptional set of measure zero, the condition can be weakened but the definitive result is not known. We shall prove the following theorem.

<u>Theorem</u>. Let $f(t) \in H_{(1/4)+\varepsilon}$, $\varepsilon > 0$. Then

$$\lim_{N\to\infty} f_N(x) = f(x) \tag{14}$$

exists a.e. On the other hand, there exists $f(t) \in H_{(1/8)-\varepsilon}$ such that

$$\overline{\lim_{N\to\infty}}\, |f_N(x)| = \infty \quad \text{a.e.}$$

For the proof of (14) we use the Kolmogorov-Seliverstov-Plessner method, based on the following lemma.

<u>Lemma</u>. Let $a, b$ be real numbers in $(-2,2)$ and suppose $0 < \alpha < 1$. Then

$$\left|\int_{-\infty}^{\infty} e^{i(a\xi+b\xi^2)}\frac{d\xi}{|\xi|^\alpha}\right| \le C_\alpha\left\{|b|^{\alpha-1/2}|a|^{-\alpha}+|a|^{\alpha-1}\right\}. \tag{15}$$

<u>Proof</u>. (Lemma) We may assume $b > 0$ and set $t = b^{1/2}\xi$ and $2A = ab^{-1/2}$. The integral to estimate becomes

$$b^{-1/2+\alpha/2}\int_{-\infty}^{\infty} e^{i(2At+t^2)}\,\frac{dt}{|t|^{\alpha}}\,.$$

If $|A| \le 2$, i.e., $|a| \le 4b^{1/2}$, then the integral is bounded and (15) has the estimate $Cb^{-(1/2)+\alpha/2} \le C|a|^{\alpha-1}$. We may therefore assume $2 < A < \infty$. First, we have

$$\int_{-1}^{1} e^{i(2At+t^2)}\,\frac{dt}{|t|^{\alpha}} \le CA^{\alpha-1} = C|a|^{\alpha-1}\cdot b^{1/2-\alpha/2},$$

which is sufficient. The integral over $(1,\infty)$ gives the stronger estimate $O(A^{-1})$. Finally, decompose the integral over $(-\infty,-1)$ into $(-\infty,-2A)$, $(-A/2,-1)$, and $(-2A,-A/2)$. Since

$$\left|\frac{d}{dt}(2At+t^2)\right| \ge A$$

the first two integrals have bounds $O(A^{-1})$. The third finally is written

$$\int_{-A/2}^{A} \frac{e^{iu^2}}{(A+u)^{\alpha}}\,du = O(A^{-\alpha}) = O(|a|^{-\alpha}|b|^{\alpha/2})$$

and the lemma is proved.

<u>Proof</u>. (Theorem) It is sufficient to get an à-priori estimate of the maximal $f_N(x)$ in case $f \in C_0^{\infty}$. We make $N$ into a function of $x$, $N(x)$, and set $\delta(x) = \frac{1}{4N(x)}$. Using (13) we find

$$\int_{-1}^{+1} f_{N(x)}(x)dx = \int_{-\infty}^{\infty}\hat{f}(\xi)\int_{-1}^{+1} e^{ix\xi+i\delta(x)\xi^2}\,dxd\xi$$

$$\le\left(\int_{-\infty}^{\infty}|\hat{f}(\xi)|^2|\xi|^{1/2}d\xi\right)^{1/2}\left(\int_{-\infty}^{\infty}\frac{d\xi}{|\xi|^{1/2}}\left|\int_{-1}^{+1}e^{ix\xi+i\delta(x)\xi^2}dx\right|^2\right)^{1/2}.$$

The first integral is bounded by the assumption $f \in H_{(1/4)+\varepsilon}$. We write the square inside the second integral as a double integral and obtain the estimate

$$\int_{-1}^{+1}\int_{-1}^{+1}dxdy\int_{-\infty}^{\infty} e^{i\xi(x-y)+i\xi^2(\delta(x)-\delta(y))}\,\frac{d\xi}{|\xi|^{1/2}} \le$$

$$\le C_{\alpha}\int_{-1}^{+1}\int_{-1}^{+1}\frac{dxdy}{|x-y|^{1/2}} \le C$$

by the lemma. The theorem is proved.

To prove the converse we make the following construction.

Let $M \simeq \sqrt{N}$ be a large integer and consider an interval $\omega = (\alpha, \alpha+M^{-1})$. We wish to study the function

$$G(x) = \sqrt{N} \int_\omega e^{iN\lambda(x-t)^2} g(t)dt.$$

If $t = \alpha + \tau M^{-1}$ this can be written as

(16) $$e^{i\lambda N(x-\alpha)^2} \int_0^1 e^{-i2\lambda M(x-\alpha)\tau + i\lambda\tau^2} g(\alpha+\tfrac{\tau}{M})d\tau.$$

Let us choose

$$g(\alpha+\tfrac{\tau}{M}) = e^{-2i\theta\tau} \cdot e^{2iM\tau} \phi(\tau)$$

where $\phi \in C_0^\infty(\frac{1}{4},\frac{3}{4})$ and is such that

$$\left|\int_0^1 \phi(\tau)e^{iu\tau} e^{i\lambda\tau^2} d\tau\right| \geq 1, \quad |u| \leq 8, |\lambda| < 2.$$

Hence, if

(17) $$|\lambda(x-\alpha) + \tfrac{\theta}{M} - 1| \leq \tfrac{4}{M},$$

the integral (16) is of absolute value $\geq 1$.

Let us now consider $\alpha = \nu/M$, $\nu = 0,1,\ldots,\sqrt{M}$. For each $\nu$ choose $\theta = \theta_\nu = \nu$. Then for $\lambda = 1 + k/M$ (17) holds if

$$\left|x - \frac{\nu}{M} - \frac{1-\frac{\nu}{M}}{1+\frac{k}{M}}\right| < \frac{3}{M}, |k| \leq \sqrt{M},$$

i.e., since $k\nu \leq M$, if

$$\left|x - 1 + \frac{k}{M}\right| \leq \frac{2}{M}.$$

We conclude that the functions $G_\nu(x)$ constructed in this way are, for suitable choice of $\lambda = \lambda(x)$ independent of $\nu$, $\geq 1$ for any $x$ in $|x-1| \leq M^{-1/2}$. We can translate the construction by multiples of $M^{-1/2}$ and obtain $M$ functions $G_\nu(x)$ so that, for each $x$, $M^{1/2}$ functions are bounded below, $0 < x < 1$.

Let us now define

$$g(\cdot,\sigma) = \sum_1^M r_\nu(\sigma)\, g_\nu(\cdot)M^{-1/4+\rho}, \quad \rho > 0.$$

Let $G(x;\lambda;\sigma)$ be the corresponding transform. $r_\nu(\sigma)$ are the Rademacher functions and

$$G(x;\lambda(x);\sigma) = \sum_{\nu=1}^M r_\nu(\sigma)\, G_\nu(x;\lambda(x))M^{-1/4+\rho}.$$

Since

$$\sum |G_\nu(x;\lambda(x))|^2 M^{-1/2+2\rho} \geqq CM^{2\rho}$$

it follows that for a suitable choice of signs $r_\nu(\sigma)$, $G(x;\lambda(x);\sigma)$ is $\geq M^{\rho/2}$ except on a set of $x$'s of small measure.

It remains to prove that the corresponding function $f(t)$ - where the change of scale $\tau \to \text{const} + Mt$ has been made - belongs to $\mathrm{Lip}(\frac{1}{8} - \varepsilon)$. If $|t_1 - t_2| > M^{-1/2}$ the corresponding g-functions belong to different intervals and

$$|f(t_1)-f(t_2)| \leq 2\mathrm{Max}|g_\nu| < M^{-1/4+\rho} < |t_1-t_2|^{1/2-\varepsilon}.$$

If $t_1, t_2$ belong to the same interval then

$$|f(t_1)-f(t_2)| \leq |t_1-t_2|\mathrm{Max}|f'| \leq CN|t_1-t_2|M^{-1/4+\rho}$$

$$\leq CN^{7/8}|t_1-t_2|^{7/8+\varepsilon}|t_1-t_2|^{1/8-\varepsilon}$$

so this case is clear if $|t_1-t_2| < \frac{1}{N}$. If $|t_1-t_2| \geq \frac{1}{N}$ the estimate $|f(t)| < M^{-1/4+\rho} = N^{-1/8+\rho/2}$ suffices.

To get a divergent series we should clearly choose a rapidly increasing sequence of $N = N_j$ and add the corresponding functions multiplied by $N_j^{-\delta}$.

4. We now return to the Gibbs theory of the harmonic oscillator. We observed that the theory was essentially trivial. If however we introduce a restriction on the potential so that particles close to the origin are left free we obtain models which have interesting properties. We shall here discuss one which essentially is due to Kac.

Let $\psi(x)$ be a non-negative continuous function for $x \geq 0$ and we assume $\psi(0) = 0$. Consider the potential

$$U_N(q) = \psi\Big(\Big(\sum_{-N}^{N} q_\nu^2\Big)\Big/(2N+1)\Big)\sum_{-N}^{N} a_{\nu-\mu}q_\nu q_\mu .$$

We wish to evaluate

$$e^{-NF(\beta)} \sim \int e^{-\beta U_N(q)}\,dq.$$

We make the change of variables

$$x_j = \frac{\sqrt{2}}{2N+1}\sum_{-N}^{N} q_\nu \cos\frac{2\pi\nu j}{2N+1},$$

$$x_0 = \frac{1}{2N+1} \sum_{-N}^{N} q_\nu ,$$

and

$$x_{-j} = \frac{\sqrt{2}}{2N+1} \sum_{-N}^{N} q_\nu \sin \frac{2\pi\nu j}{2N+1} .$$

Using earlier notations and $|x|^2 = \sum_{-N}^{N} x_j^2$, we obtain

$$U_N(q) = \psi(|x|^2) \sum_{-N}^{N} A_j^{(N)} x_j^2 (2N+1),$$

where

$$A_j^{(N)} = A^{(N)}\left(\frac{2\pi j}{2N+1}\right).$$

Since the transformation is orthogonal, our problem can be formulated as follows. Given $0 \le A_1 \le A_2 \le \ldots \le A_N \le C$, we wish to estimate

$$\text{(18)} \qquad \int e^{-N\psi(|x|^2)\sum_1^N A_j x_j^2} \, dx_1 \ldots dx_N .$$

<u>Lemma</u>. Let $f(t)$ and $g(t)$ be continuous functions on $(0,1)$, increasing and decreasing, respectively. Suppose $f(0) = g(1) = 0$. Then

$$S = \sup_t f(t)g(t) \leq - \int_0^1 f(t)dg(t) \leq S \log\left(\frac{f(1)g(0)e}{S}\right).$$

<u>Proof</u>. Clearly, for all $t_0$, we have

$$f(t_0)g(t_0) = - \int_{t_0}^1 f(t_0)dg(t) \leq - \int_0^1 f(t)dg(t).$$

Now choose $t_0$ so that $g(t_0)f(1) = S$. Then

$$- \int_0^1 fdg = - \int_0^{t_0} - \int_{t_0}^1 \leq - S \int_0^{t_0} \frac{dg}{g} - f(1) \int_{t_0}^1 dg =$$

$$= - S \log\left(\frac{g(t_0)}{g(0)}\right) + f(1)g(t_0) = S + S \log\left(\frac{f(1)g(0)}{S}\right).$$

We write (18) in polar coordinates: $dx_1 \ldots dx_N = r^{N-1} dw$ and wish to evaluate the surface integral for a fixed value of $r$. Let $C$ be a large constant and write $-B_j = A_j - C$ so that $B_j > 0$. We now wish to estimate

$$\text{(19)} \qquad e^{-N\psi(r^2)Cr^2} \, r^{N-1} \int_{|x|=1} e^{N\psi(|x|^2)r^2 \sum_1^N B_j x_j^2} \, dw .$$

Let us set $\gamma = \psi(r^2)r^2$. We may clearly replace the surface integral by a volume integral over $|x| \leq 1$ since $B_j \geq 0$. Now take two numbers $\alpha'$ and $\alpha''$, $\alpha' + \alpha'' = 1$, and divide the set of indices into two groups: $0 < j \leq \alpha'N$ and $\alpha'N < j \leq N$. Write $x = (x', x'')$ and divide $\sum = \sum' + \sum''$. Let $V_N$ be the volume of the N-sphere. Then

$$V_N I_N = \int_{|x|<1} e^{\gamma N \sum B_j x_j^2}\, dx = \int e^{\gamma N \sum'} dx'\, e^{\gamma N \sum''} dx'' =$$

$$= - \int_0^1 \left( \int_{|x'|<t} e^{\gamma N \sum'} dx' \right) d\left( \int_{|x''|<\sqrt{1-t^2}} e^{\gamma N \sum''} dx'' \right)$$

$$= - \int_0^1 f(t) dg(t).$$

Clearly $f(1) \sim C^N V_{\alpha' N}$ and similarly for $g(0)$. Observe also that

$$V_N \simeq N^{-N/2} C^N.$$

In the lemma $\frac{f(1)g(0)}{S} \sim C^N N^{\beta N}$ for some $\beta$. We conclude

$$V_N I_N = \sup_\rho \int_{|x'| \leq \rho} e^{\gamma N \sum'} dx' \int_{|x''| \leq \sqrt{1-\rho^2}} e^{\gamma N \sum''} dx'' \cdot Q$$

where

$$1 \leq Q \leq CN\log N.$$

We can repeat the argument and obtain

$$V_N I_N = \sup \prod_{k=1}^{p} \int_{|x^{(k)}|<\rho_k} e^{\gamma N \sum^{(k)} B_j x_j^2} dx^{(k)}\, Q_p,$$

where

$$1 \leq Q_p \leq (CN\log N)^p$$

and

(20) $$\rho_1^2 + \rho_2^2 + \ldots + \rho_p^2 = 1.$$

Taking a suitable mean value, $B_j^*$, of $B_j$ in each block, we find

$$V_N I_N = \sup_{(\rho)} e^{\gamma N \sum_1^p B_j^* \rho_j^2 + \alpha_j N \log \rho_j} \prod_1^p V_{\alpha_j N}\, Q_p.$$

We can compute the maximum and obtain

(21)
$$\begin{cases} 2\gamma B_j^* \rho_j + \dfrac{\alpha_j}{\rho_j} = K\rho_j \\[2ex] \rho_j^2 = \dfrac{\alpha_j}{K-2\gamma B_j^*} \end{cases}$$

for extreme values, where $K \geqq 2\gamma B_j^*$ for all $j$ and (20) gives $K$ uniquely.

We can now first let $N \to \infty$ and then $P \to \infty$, and let Max $(\alpha_{j+1}-\alpha_j) \to 0$. In this way we find, for $B(x) = C-A(x)$, that

$$\lim_{N\to\infty} \frac{\log I_N}{N} = \gamma \frac{1}{2\pi}\int_{-\pi}^{\pi} \frac{B(x)}{K-2rB(x)}\,dx - \frac{1}{2\pi}\int_0^{2\pi} \log(K-2rB(x))dx$$

provided that

$$\frac{1}{2\pi}\int_{-\pi}^{\pi} \frac{dx}{K-2rB(x)} = 1.$$

Going back to the original variables, the logarithm of (19) is asymptotic to $N$ times

$$- Cr^2\psi(r^2) + \log r + Cr^2\psi(r^2) - \frac{1}{2\pi}\int_{-\pi}^{\pi} \frac{A(x)}{\lambda+2A(x)} - \frac{1}{2\pi}\int_{-\pi}^{\pi} \log(2A(x)+\lambda) - \log(\psi(r^2)r^2),$$

where $\lambda \geqq 0$ satisfies

(22)
$$\frac{1}{2\pi}\int_{-\pi}^{\pi} \frac{dx}{2A(x)-\lambda} = r^2\psi(r)^2.$$

$\lambda$ is a function of $r$ and to find (18) we must take

(23)
$$\sup_r \left\{ - \frac{1}{2\pi}\int_{-\pi}^{\pi} \frac{A(x)}{\lambda+2A(x)}\,dx - \frac{1}{2\pi}\int_{-\pi}^{\pi} \log(2A(x)+\lambda)dx - \log((\psi(r^2))r^2) \right\}.$$

It is clear that the dependence of a parameter $\beta$ in front of $A$ can be extremely complicated and that different types of singularities can occur. We also note that the whole discussion depends on whether (22) has a solution or not, i.e., whether or not

(24)
$$\frac{1}{2\pi}\int_{-\pi}^{\pi} \frac{dx}{2A(x)} \geqq r^2\psi(r^2).$$

Let us consider the case studied by Kac, $\psi(r) = 0$, $r \leq 1$, $\psi(r) = 1$, $r > 1$, and assume first

(25)
$$\frac{1}{2\pi}\int_{-\pi}^{\pi} \frac{dx}{2A(x)} \geq 1.$$

Differentiating (23) and using (22) we see that the maximum in (23) is taken for $\lambda = 0$ and the whole expression becomes $= C \log \beta$ if $A(x)$ is replaced by $\beta A(x)$. In terms of our time evolution this means that if (25) holds then the fact that $U = 0$ for $\sum q_\nu^2 < 2N + 1$ does not interfere with the motion, because it does not take place in this region.

If, however, (25) does not hold, the limiting procedure of (21) changes. In the limit it must hold that

$$\sum_0^{\delta N} x_j^2 \geq C > 0, \quad \text{as first } N \to \infty, \text{ then } \delta \to 0.$$

Going back to our original variables this means that for all $m$

$$E\left(\left(\frac{q_1+\ldots+q_m}{m}\right)^2\right) \geq C > 0,$$

where $E$ is expectation with respect to the Gibbs distribution. The free energy can easily be computed explicitly in this case also.

In terms of the time evolution, the movement will take place close to the sphere $\sum q_\nu^2 = 2N + 1$ and either move freely or get trapped in those parts of the sphere where the potential $U(q)$ is small.

## III. <u>One-dimensional Models and Markov Chains</u>.

We shall in this chapter consider models on the real line where the interaction between successive particles is increasingly more and more dependent. We begin with a model usually called van Hove's. Here the potential is assumed to have homogeneously long range which is compensated by an assumption of small forces.

Let $\Phi(x)$ be a continuous differentiable function on the torus $\mathbb{T}$ of length 1. We assume that $\Phi(x)$ is even and has mean value zero:

$$\Phi(x) = \sum_1^\infty c_n \cos 2\pi nx.$$

Hence, we have

$$\frac{1}{N} \sum_{\nu,\mu=1} \Phi(x_\nu - x_\mu) = N \sum_{n=1}^\infty c_n \left| \frac{1}{N} \sum_{\nu=1}^N e^{2\pi i n x_\nu} \right|^2 .$$

We are interested in the asymptotic behavior of

$$(1) \qquad e^{-Nf_N(\beta)} = \int_0^1 \cdots \int_0^1 e^{-\frac{\beta}{N}\sum \Phi(x_\nu - x_\mu)} dx_1 \ldots dx_N .$$

We now choose sets $E_N \subset \mathbb{T}^N$ so that

(2) $$\frac{1}{N}\sum_{1}^{N} e^{2\pi i n x_\nu} \to \int_0^1 e^{2\pi i n x}\, d\sigma(x) = \hat{\sigma}(n)$$

where $\sigma \geqq 0$, $\int_0^1 d\sigma = 1$, and

(3) $$e^{-N\beta\sum_1^\infty c_n|\hat{\sigma}(n)|^2}\, mE_N = e^{-Nf_N(\beta)+o(N)}.$$

This is clearly always possible. We use the following

Lemma. Divide the interval $(0,1)$ into $k$ equal intervals $I_j$ and suppose that $a_j N$ $x_\nu$'s belong to $I_j$. This defines a set $E(a_1,\ldots,a_k)$. Then

$$mE(a_1,\ldots,a_k) = \exp\left(\{-\sum_1^k a_j \log a_j - \log k\}N + o(N)\right).$$

Proof.

$$mE = \frac{N!}{(a_1N)!\ldots(a_kN)!}\, k^{-N} \sim \prod_1^k a_j^{-a_jN} \cdot k^{-N}.$$

To continue our discussion of (1) choose $a_j = \sigma(I_j)$. In the limit (2) the corresponding number of $x_\nu$'s in each interval varies by at most $\pm\varepsilon N$. Totally this gives $<N^k$ choices of $a_j$. It follows that for all $\varepsilon$

$$mE_N \leq \exp\{(-\sum a_j \log(a_j k) + \varepsilon)N\}$$

for $N$ large enough. A converse inequality also follows from the lemma. Hence, if we define $\phi_k(x) = a_j k$, $x \in I_j$, it follows that

$$\int_0^1 \phi_k(x)dx = 1,\ \phi_k dx \to d\sigma \text{ weakly},$$

and

$$\exp\{-N\int_0^1 \phi_k \log \phi_k\, dx\} = mE_N e^{o(N)}, \qquad N \to \infty \text{ and } k \to \infty.$$

The result can be formulated as follows. Let $d\sigma$ be a probability measure. Consider all choices of points $x_\nu$ in $\mathbb{T}$, $\nu = 1,\ldots N$, whose distribution lies in some weak $\varepsilon$-neighborhood of $d\sigma$. This defines a set $E_N$. Then, as first $N \to \infty$, then $\varepsilon \to 0$, we have

$$mE_N = \exp\left\{-N\int \left(\frac{d\sigma}{dt}\log\frac{d\sigma}{dt}\right)dt + o(N)\right\}$$

with the interpretation that if the integral diverges (in particular if $\sigma$ is non-absolutely continuous) $\dot{m}E_N < \exp(-AN)$ for all $A < \infty$.

Going back to $f(\beta)$ we find

$$-\varliminf_{N\to\infty} f_N(\beta) \leq \varliminf_{k\to\infty}\left[-\beta\int_0^1\int_0^1 \Phi(x-y)\phi_k(x)\phi_k(y)dxdy - \int_0^1 \phi_k\log\phi_k dx\right].$$

On the other hand, let $\psi \geq 0$ be arbitrary with $\int_0^1 \psi dx = 1$. Then

$$e^{-Nf_N(\beta)} = \int_0^1\cdots\int_0^1 e^{-\frac{\beta}{N}\sum_{\nu,\mu=1}^{N}\Phi(x_\nu-x_\mu)-\sum_{\nu=1}^{N}\log\psi(x_\nu)}\prod_1^N\psi(x_\nu)dx_1\cdots dx_N$$

$$\geq \exp\left\{-\frac{\beta}{N}\int_0^1\cdots\int_0^1\sum_{\nu,\mu=1}^{N}\Phi(x_\nu-x_\mu)\prod_1^N\psi(x_\nu)dx_1\cdots dx_N - \int_0^1\cdots\int_0^1 -\sum_{\nu=1}^{N}\log\psi(x_\nu)\prod_1^N\psi(x_\nu)dx_1\cdots dx_N\right\}$$

$$\geq \exp\left\{-\frac{\beta}{N}(N-1)N\int_0^1\int_0^1\Phi(x-y)\psi(x)\psi(y)dxdy - N\int_0^1\psi\log\psi dx + O(1)\right\}.$$

This means that

$$-\overline{\lim}\, f_N(\beta) \geq -\int_0^1\int_0^1\beta\Phi(x-y)\psi(x)\psi(y)dxdy - \int_0^1\psi\log\psi dx$$

for all $\psi$ and hence $f = \lim f_N(\beta)$ exists and

(4) $$f(\beta) = \inf_{\psi}\left\{\int_0^1\int_0^1\beta\Phi(x-y)\psi(x)\psi(y)dxdy + \int_0^1\psi\log\psi dx\right\}$$

where $\psi \geq 0$ and $\int_0^1 \psi dx = 1$.

The functional in (4) will be designated by $I(\psi)$.

<u>Lemma</u>. There exists $\phi$ minimizing (4); $\phi$ is continuous and positive and satisfies the non-linear equation

(5) $$2\beta\Phi*\phi + \log\phi = C = \text{Constant}.$$

<u>Proof</u>. Let $\psi_n$ be a minimizing sequence. Clearly, we have

$$\int_0^1 \psi_n \log^+ \psi_n dx < \text{Const}.$$

Hence $\psi_n$ is uniformly integrable. The weak limits are therefore non-singular. Let $\phi$ be such a weak limit. By Fatou's lemma $\phi$ is minimizing. Let, for some $a > 0$, $E_a$ be the set where $\phi(x) > a$.

Take $\psi$ with support on $E_a$, such that $\psi \in L^\infty$ and $\int_0^1 \psi dx = 1$. Then

$$I(\phi) \leq I(\phi+\delta\psi) = I(\phi) + 2\delta \int_{E_a} \beta\Phi*\phi\psi dx + \delta \int_{E_a} (\log\phi)\psi dx + O(\delta^2).$$

Hence, we have

$$2\beta\Phi*\phi + \log\phi = c \quad \text{on} \quad \{x|\phi(x)>0\}.$$

Since the first term is bounded, $\phi(x) \geq a > 0$ if $\phi(x) > 0$. Suppose finally that $\phi = 0$ on $E_0$ with $mE_0 > 0$. Let $\psi_0$ be the characteristic function of $E_0$ and consider

$$\phi_\delta = (1-\delta)\phi + b\delta\psi_0, \quad \text{where } \delta > 0 \text{ and } b = (mE_0)^{-1}.$$

Then

$$I(\phi_\delta) = I(\phi) + O(\delta) + \delta\log\delta < I(\phi)$$

for $\delta$ small enough. The lemma is therefore proved.

Let us now discuss the function $f(\beta)$, and assume the normalization $|\Phi(x)| \leq 1$. By (5), we have

$$|\log\phi - c| \leq 2\beta,$$

i.e.,

$$e^c e^{-2\beta} \leq \phi(x) \leq e^c e^{2\beta}.$$

Hence

$$|c| \leq 2\beta$$

and

$$e^{-4\beta} \leq \phi(x) \leq e^{4\beta}.$$

Let $\phi(x_a) = \operatorname{Max}\phi(x) = 1 + a$, $\phi(x_b) = \operatorname{Min}\phi(x) = 1-b$, $a,b \geq 0$. From (5) it follows that

$$\log\frac{1+a}{1-b} = 2\beta \int_0^1 (\phi(x_a-t) - \phi(x_b-t))\Phi(t)dt \leq 2\beta(a+b),$$

i.e.,

$$\log(1+a) - 2\beta a \leq \log(1-b) + 2\beta b \leq 0$$

if $\beta \leq \frac{1}{2}$. If $\beta \leq \frac{1}{4}$ then $a \leq e - 1$ and

$$\log(1+(e-1)) - \tfrac{1}{2}(e-1) > 0.$$

We conclude that $a = b = 0$ if $\beta \leq \frac{1}{4}$, and hence $f(\beta) \equiv 0$, $\beta \leq \frac{1}{4}$. The following theorem holds.

<u>Theorem</u>. If $|\Phi(x)| \leq 1$ then

$$f(\beta) \equiv 0 \quad \text{for} \quad 0 < \beta < \tfrac{1}{4}.$$

If $\Phi(x)$ is positive definite, i.e., $c_n \geqq 0$ then $f(\beta) \equiv 0$, $0 < \beta < \infty$. If $\Phi$ is not positive definite then $f(\beta) < 0$ for $\beta$ large.

We have proved the theorem except for the last two statements. We observe that for any $\psi \geq 0$, $\int_0^1 \psi dx = 1$ and

$$\int_0^1 \psi \log\psi dx \geqq 0.$$

This follows from our equation (5) with $\Phi \equiv 0$. Hence, if $\Phi$ is positive definite, $I(\psi) \geqq 0$ and so $f \equiv 0$. Assume therefore $c_n < 0$ and choose

$$\psi = 1 + \cos 2\pi nx.$$

Hence,

$$f \leq \beta c_n \frac{1}{4} + \int_0^1 (1+\cos 2\pi nx)\log(1+\cos 2\pi nx)dx$$

$$< 0 \quad \text{for } \beta \text{ large enough.}$$

Remark. This system with $\Phi$ non-positive-definite shows a "phase-transition", i.e., $f(\beta) \not\equiv$ analytic function in $\beta$. More precise information on the nature of $f(\beta)$ can however be deduced from the above discussion. In a forthcoming thesis, M. Tamm proves that $(\beta)$ is analytic except at a finite number of points $\beta$.*

An interesting question here is the study of the time-evolution. If we e.g. assume that all particles start in $x = 0$ we obtain a non-linear equation for the distribution of particles at time $t$ in terms of the distribution of velocities at $t = 0$. What is the condition on this initial distribution in order that Gibbs' theory holds?

In the discussion of the van Hove potential, we need only study the distribution of points without any coupling between points.

Let us now first consider the following (trivial) problem. Let $A_{ij}$ be a $k \times k$ square matrix with positive entries. We wish to study the asymptotic behavior of

$$S = \sum_{(i_1 \ldots i_N)} A_{i_1 i_2} A_{i_2 i_3} \ldots A_{i_{N-1} i_N} \quad \text{as } N \to \infty.$$

Exactly as in the previous case the main contribution must come from some distribution of pairs of indices where

$$\text{(6)} \qquad \sum p_i = 1, \quad \sum_j m_{ij} = 1$$

and

$$\text{(7)} \quad S = \sup_{(m_{ij})} e^{N\left[\sum_{i,j} (\log A_{ij}) p_i m_{ij} - \sum_{i,j} p_i m_{ij} \log m_{ij}\right] + o(N)}.$$

*If $f$ also depends on a density $\rho$, then the singular set in $(\beta,\rho)$ is a finite union of analytic curves.

$m_{ij}$ are the transition probabilities of a Markov process and $p_i$ are the probabilities of $i$, so that $p_i m_{ij}$ is the frequency of $(i,j)$.

We can now study this as a variational problem. Let $p_i$ be fixed and vary $m_{ij}$ by small quantities $\mu_{ij}$ so that

$$\sum_{j=1}^{k} \mu_{ij} = 0, \quad i = 1\ldots k$$

and

$$\sum_{i=1}^{k} p_i \mu_{ij} = 0, \quad j = 1\ldots k.$$

The variational equations are

$$0 = \sum_{i,j} (\log A_{ij}) p_i \mu_{ij} - \sum_{i,j} p_i \mu_{ij} \log m_{ij}.$$

We obtain

(8) $$m_{ij} = x_i y_j A_{ij}.$$

We can also make a small variation $q_i$ of $p_i$ so that

$$\sum_{i=1}^{k} (q_i m_{ij} + p_i \mu_{ij}) = q_j, \quad j = 1,\ldots,k.$$

We find

$$U = \sum \log A_{ij}(q_i m_{ij} + p_i \mu_{ij}) - \sum q_i m_{ij} \log m_{ij} - \sum p_i \mu_{ij} \log m_{ij}$$

$$= - \sum (\log x_i + \log y_i) q_i.$$

We can vary $q_i$ freely, $\sum q_i = 0$, by choosing $\mu_{ij}$, and hence

$$m_{ij} = K^{-1} \frac{x_i}{x_j} A_{ij}.$$

We obtain

$$\sum_{i=1}^{k} x_i A_{ij} = K x_j$$

and the minimum value = $\log K$. If $A_{ij}$ is assumed symmetric we also obtain

$$p_i = x_i^2$$

and $K$ is the largest eigenvalue of $A_{ij}$.

The result is now obvious. To estimate $S$ study instead

$$S' = \sum_{(i)} A_{i_1 i_2} \cdots A_{i_{N-1} i_N} x_{i_N} = K^N \sum_{i=1}^{k} x_i,$$

and clearly $S$ and $S'$ are comparable.

In analogy with the above discussion let us prove the following theorem due to Beurling; there is related work by Jamison.

Theorem. Let $K(x,y) > 0$ be continuous and symmetric on $(0,1)$. Then there is a unique solution $f(x) > 0$ to the equation

$$f(x) = \int_0^1 \frac{K(x,y)}{f(y)}\, dy.$$

Proof. To get the existence define, for $\phi > 0$,

$$I(\phi) = \int \log\phi(x)dx - \frac{1}{2}\iint K(x,y)\phi(x)\phi(y)dxdy.$$

It is easy to see as before that $\sup_\phi I(\phi)$ has a solution $\phi_0 > 0$.

$$\delta^{-1}(I(\phi_0+\psi\delta) - I(\phi_0)) = \int \frac{\psi}{\phi_0}\, dx - \int \psi(x)dx \int K(x,y)\phi_0(y)dy + o(1),$$

so $f = 1/\phi_0$ is a solution.

Suppose now that $f$ and $g$ are solutions. Then

$$0 = \iint K(x,y) \left[\frac{1}{f(x)f(y)} - \frac{1}{g(x)g(y)}\right]\left[\frac{f(x)}{g(x)} - \frac{g(y)}{f(y)}\right] dxdy$$

since

$$\int_0^1 \frac{f(x)}{g(x)}\, dx = \int_0^1 \frac{g(x)}{f(x)} = \iint \frac{K(x,y)}{f(x)g(y)}\, dxdy.$$

Hence,

$$0 = \iint K(x,y) \left[\frac{f(x)}{g(x)} - \frac{g(y)}{f(y)}\right]^2 \frac{1}{f(x)g(y)}\, dxdy,$$

and so

$$\frac{f(x)}{g(x)} \equiv \frac{g(y)}{f(y)} \ (= 1).$$

Remark. The above result leads to the following problem, which is a continuous version of the continued fraction expansion. Let $h_0(x) > 0$ and continuous on $(0,1)$. Form

$$h_n(x) = \int_0^1 \frac{K(x,y)}{h_{n-1}(y)}\, dy, \quad n = 1,2,\ldots .$$

When does

$$\lim_{n\to\infty} h_n(x) (= f(x)) \text{ exist?}$$

Our goal is to study the partition function $S$ for general poten tials. This problem is still unsolved. We shall here consider two special cases:

1. $\Phi(x) = e^{-|x|}$

2. $\Phi(x) > 0$ with compact support.

The first case has been treated several times earlier. The present direct method may have a certain interest.

To study 1 observe that we may write

$$\int_0^L \cdots \int_0^L e^{-\beta \sum e^{-|x_\nu - x_\mu|}} dx_1 \cdots dx_N$$

$$= N! \int_{0<x_1<\cdots<x_N<L} \cdots \int e^{-\beta e^{x_1}(e^{-x_2}+\cdots+e^{-x_N}) - \beta e^{x_2}(e^{-x_3}+\cdots+e^{-x_N}) - \cdots} dx_1 \cdots dx_N$$

Introduce new coordinates

$$e^{-y_i} = \sum_i^N e^{-x_j}, \quad e^{-x_i} = e^{-y_i} - e^{-y_{i+1}}, \quad \text{and} \quad y_{N+1} = 0 .$$

The integral becomes

$$\int_{D_y} \cdots \int \exp -\beta \left\{ \frac{e^{-y_2}}{e^{-y_1} - e^{-y_2}} + \cdots + \frac{e^{-y_N}}{e^{-y_{N+1}} - e^{-y_N}} \right\}$$

$$\times \prod_1^N [e^{y_i}(e^{-y_i} - e^{-y_{i+1}})]^{-1} \, dy_i \cdots dy_N .$$

Writing $t_i = y_{i+1} - y_i$, the integral is

$$\int_{D_y} \cdots \int e^{\sum F(t_i)} dt_i \cdots dt_N .$$

if

$$F(t) = -\frac{\beta}{e^t - 1} - \log(1 - e^{-t}), \qquad t > 0. \tag{9}$$

The original condition $0 < x_1 < \ldots < x_N < L$ has become

$$D_t : \begin{cases} t_i \geqq 0 \\ e^{t_i} + e^{-t_{i+1}} \geqq 2 \\ \sum_1^N t_i \leqq L. \end{cases}$$

There is some small error near $x = 0$ which is easily checked to make no difference. We have therefore a problem of the type studied earlier with the Markov condition in the domain. It has the following form.

Let $m(x,y)$ be transition probabilities for a Markov Chain so that

$$m(x,y) = 0, \ (x,y) \notin M.$$

Let $p(x)$ be the corresponding density and assume

$$\int x\, p(x)dx \le \rho^{-1}. \tag{10}$$

Solve the variational problem

$$\sup_{(m)}\left[\int p(x)F(x)dx - \iint p(x)m(x,y)\log m(x,y)dxdy\right].$$

In our case $F$ was given by (9); $M$ is defined in $D_t$ and $\rho = N/L$.

We consider instead the finite problem (7) where now $A_{ij} = F_i$ for all $j$. The variational result (8) here yields

$$m_{ij} = m_j c_i \quad , \quad c_i = \Big(\sum_{M_i} m_j\Big)^{-1}$$

where $M_i = \{j \mid (i,j) \in M\}$.

The second type of variation yields

$$F_i - \log c_i - \log m_i = a' + b' i$$

if we also take (10) into consideration.

The continuous version of this is

$$F(x) - \log\left(\frac{m(x)}{\int_{E_x} m(t)dt}\right) = a + bx$$

where $E_x = \{t \mid (x,t) \in M\}$. Writing

$$M(x) = \int_{E_x} m(t)dt$$

our transition matrix $m(x,y) = \dfrac{m(y)}{M(x)}$, $y \in E_x$, is determined by

$$\frac{m(x)}{M(x)} = Ae^{F(x)}\, e^{-bx}. \tag{11}$$

$A$ and $b$ are to be determined from the conditions, $\int_0^\infty p(x)dx = 1$

and $\int_0^\infty xp(x)dx = \rho^{-1}$. It is interesting to note how (11) determines $m(x)$. First, if $x \geq \log 2$, then $E_x = (0,\infty)$ and $m(x)$ is computed. Then, iteratively, $m(x)$ is determined in the intervals $(x_n, x_{n-1})$ where

$$e^{x_n} + e^{-x_{n-1}} = 2, \quad x_0 = \log 2.$$

One sees that

$$(12) \qquad x_n = \log \frac{n+2}{n+1}, \quad x_n \to 0, \ n \to \infty.$$

To compute $p(x)$ we set $\pi(x) = p(x)/m(x)$. If $\psi(y)$ is determined by $e^{\psi(y)} + e^{-y} = 2$, then $\psi(y) < y$ and

$$\pi(y) = \int_{\psi(y)}^\infty \pi(x) A e^{F(x)} e^{-bx}\, dx = C - \int_0^{\psi(y)} .$$

This equation is solved easily by iterations. The solutions depend analytically on $\beta, A$ and $b$. Whether or not the free energy depends analytically on $\beta$ and $\rho$ for all choices of $(\beta,\rho)$ is not clear. It should be noted that the model shows discontinuities in the derivatives of $p$ for the values $x_n$ in (12). These correspond to certain quotients of the sums of the original variables.

Let us now consider the case 2 above. We assume that $\Phi(x)$ has support in $(-\frac{1}{2},\frac{1}{2})$ and that $\Phi(x) > 0$. We wish to compute asymptotically

$$S = \int_{-L}^{L} \cdots \int_{-L}^{L} e^{-\beta \sum \Phi(x_\nu - x_\mu)}\, dx_1 \ldots dx_N, \quad M = 2L+1.$$

Let $I_j$ be the interval $(j,j+1)$. Let $a_j$ be the number of points $x_\nu \in I_j$. We can then write $S$ as

$$(13) \qquad S = N! \sum_{\substack{a_{-M},\ldots,a_M=0 \\ a_{-M}+\ldots+a_M=N}}^{N} \frac{1}{a_{-M}!\ldots a_M!} \int_{\substack{x_{\nu j}\in I_j \\ \nu=1,\ldots,a_j}} \cdots \int e^{-\beta \sum \Phi(x_{\nu j} - x_{\mu k})} \prod_{\nu,j} dx_{\nu j}.$$

Let us introduce the notation $X_j = (a_j; x_{1j}, \ldots, x_{a_j j})$ and $dX_j = \frac{e^{-1}}{a_j!} dx_{1j} \ldots dx_{a_j j}$. Here, $a_j = 0,1,\ldots,$ and the $x_{\nu j}$ move in $I_j$.

We also write

$$(14) \qquad K(X_{j-1}, X_{j+1}) = e^{-\frac{1}{2}\beta \sum_{j-1}^{0}} \int e^{-\beta \sum_j} dX_j \, e^{-\frac{1}{2}\beta \sum_{j+1}^{0}},$$

where we have fixed the variables in $X_{j-1}$ and $X_{j+1}$ corresponding to the intervals $I_{j-1}$ and $I_{j+1}$. $\sum_j^0$ indicates that we only sum over variables in the same interval while $\sum_j$ is the sum over all variables in $I_j$ and all other variables (only in $I_{j-1}$ and $I_{j+1}$ by assumption concerning the range of $\Phi$). Disregarding the factor $N!e^L$, the expression to estimate is

(15) $$\int\cdots\int \prod_{j=1}^{L} K(X_j,X_{j+1})dX_1\cdots dX_L.$$

The dependence of $\beta$ is given in (14). Note that $K \leq 1$ for $\beta \geq 0$. Let

$$Q(X,Y) = \frac{1}{\sqrt{a(X)}\sqrt{a(Y)}} K(X,Y),\ a \geqq 1,$$

and let $\lambda$ be the largest eigenvalue of

(16) $$\int Q(X,Y)f(Y)a(Y)dY = \lambda f(X).$$

If we rewrite (15) using $Q$ and replace the last integration by $f(X_L)dX_L$ we see that (15) is $\leq C\lambda^L$. On the other hand, $f(X) \geq e^{-ca(x)}$ and

$$\int e^{ca(X)}\, dX < \infty,$$

which gives an estimate in the opposite direction. We conclude that $\lambda^L$ is the asymptotic behavior. Since the largest eigenvalue is isolated by the following lemma, $\lambda$ depends analytically on the parameter $\beta$.

Lemma. Let $K(x,y)$ be symmetric, $0 \leq K(x,y) \leq 1$ on $(0,1)$. Suppose that $K(x,y) \geq \delta > 0$ except on a set of measure $< \varepsilon(\delta)$, $\varepsilon(\delta) \to 0$, $\delta \to 0$. Let $\lambda_0$ be the largest eigenvalue. Then there is an $r$ depending only on $\varepsilon(\delta)$ so that $|z-\lambda_0| < r$ contains no other eigenvalue.

Proof. Let $f \geq 0$ be an eigenfunction corresponding to $\lambda_0$, let $\phi(x)$ correspond to $\mu$, and assume $|\mu| \geq \frac{1}{2}\lambda_0$. It is easy to see that.

$$q = \iint K(x,y)dxdy,\quad q \leq \lambda_0 \leq 1,\quad 0 < f < \frac{1}{q},\ \text{and}\ |\phi| < \frac{2}{q}.$$

Clearly, $q \geq \delta(1-\varepsilon(\delta))$. For $x \notin E$ and $mE < \sqrt{\varepsilon}$, $K(x,y) \geq \delta$ except on a set of measure $<\sqrt{\varepsilon}$ . Hence, if $x \notin E$ then

$$\lambda_0 f(x) \geq \delta\left(\int_0^1 f(y)dy - \frac{1}{q}\sqrt{\varepsilon}\right) \geq \delta\,\frac{1-\sqrt{\varepsilon}}{q}\,.$$

Hence $f(x)$ is bounded below except in a small set. Furthermore,

$$\int_0^1 f(x)\phi(x)dx = 0 \;,\; \int_0^1 \phi(x)^2\,dx = 1.$$

It follows that

$$\int f|\phi|dx \geqq \delta \int_{f>\delta} |\phi|dx \geqq \frac{\delta q}{2}\int_{f>\delta} \phi^2\,dx \geqq \frac{\delta q}{2}(1-\delta'),$$

and so

$$\int_{\phi<0} |\phi|dx \geqq \delta'' \;,\; \int_{\phi>0} |\phi|dx \geq \delta''.$$

Hence, we have

$$\begin{aligned} |\mu| &= \left|\iint K(x,y)\phi(x)\phi(y)dxdy\right| \leqq \\ &\leqq \iint K(x,y)|\phi(x)\phi(y)|dxdy - \delta''' \\ &\leq \lambda_0 - \delta'''. \end{aligned}$$

We have therefore proved the following theorem.

<u>Theorem</u>. If $\Phi(x) \geq 0$ is continuous with compact support then the free energy is an analytic function of $\beta$.

The reasonable assumption on $\Phi(x)$ for the validity of the result is that $\Phi$ is positive definite. However, in the above approach we only get a bound,

$$K(X,Y) \leq e^{Ca(X)a(Y)},$$

and

$$\sum \frac{e^{\nu\mu}}{\nu!\mu!} \text{ diverges.}$$

However, if we also introduce changes of signs in a random way so that the system is neutral, then the conclusion still holds.

We consider the following set-up. Let $\Phi(x)$ be positive definite and of compact support. Let $\mu$ be an even probability measure. Define

$$I_N = \frac{1}{N!}\int\cdots\int d\mu(\varepsilon_1)\ldots d\mu(\varepsilon_N)\int_{0<x_\nu<N} e^{-\beta\sum\varepsilon_\nu\varepsilon_\mu\Phi(x_\nu-x_\mu)}dx_1\ldots dx_N\,.$$

Theorem.

$$\lim_{N\to\infty} \frac{\log I_N}{N} = F(\beta)$$

exists and is analytic for $0 < \beta < \infty$.

Proof. The proof depends on a reformulation of the limit, using Gaussian variables and Fourier transforms. Define

$$\Phi_N(x) = \Phi(Nx) \ , \ |x| < \frac{c}{N}$$

$$= 0 \ , \ \frac{c}{N} < |x| < \pi .$$

Then, we have

$$\Phi_N(x) = \sum_0^\infty c_n \cos nx$$

where

$$c_n = \frac{1}{N}\hat{\Phi}(\frac{n}{N}).$$

In $I_N$ we change $x_\nu \to Nx_\nu$ and introduce the Fourier series. We obtain (deleting the parameter $\beta$)

$$I_N = \frac{N^N}{N!}\int d\mu(\varepsilon)\int e^{-\sum\limits_0^\infty c_n[(\sum\limits_1^N \varepsilon_\nu \cos nx_\nu)^2+(\sum\limits_1^N \varepsilon_\nu \sin nx_\nu)^2]} dx.$$

We use the formula

$$e^{-ct^2} = \frac{1}{\sqrt{2\pi}}\int_{-\infty}^{\infty} e^{-\frac{\xi^2}{4}+itc\xi} d\xi$$

and introduce one variable $\xi = \xi_n$ for each $c = c_n$, $n = 0,\pm1,\pm2,\ldots$, and each of the two cosine and sine sums in the exponent. If we define $\gamma_n^2 = c_n$, $\gamma_n \geq 0$, and observe that $\gamma_n$ is rapidly decreasing, we can set

$$F_N(x;\xi) = \sum_0^\infty \gamma_n(\xi_n \cos nx+\xi_{-n}\sin nx).$$

After some computations we obtain

$$I_N = \lim_{M\to\infty} \frac{N^N}{N!}\left(\frac{1}{2\pi}\right)^{M+\frac{1}{2}} \int\limits_{\mathbb{R}^{2M+1}} e^{-\frac{|\xi|^2}{4}} d\xi \left(\frac{1}{2\pi}\int_0^{2\pi} \hat{\mu}(F_N(x;\xi))dx\right)^N .$$

Let us now write

$$\frac{1}{2\pi}\int_0^{2\pi} \hat{\mu}(F_N)dx = \sum_1^{N/A} \frac{1}{2\pi}\int_{\omega_j} \hat{\mu}(F_N)dx = \frac{1}{N}\sum_1^{N/A} NX_j ,$$

where $\omega_j$ is an interval of length $\frac{2\pi A}{N}$. If, (a), $NX_j$ were independent with distribution $X(A)$, and if, (b), $X(A)$ was independent of $N$, then we could use the computation of high moments in Chapter I to obtain

$$E\left[\left(\frac{1}{2\pi}\int \hat{\mu}(F_N)dx\right)^N\right] \sim \exp\{N(\psi_A(\lambda) - 1 - \log \lambda)\}$$

where

$$(17) \qquad \psi_A(\lambda) = \frac{1}{A} \log E(e^{\lambda X(A)}).$$

and

$$\lambda\psi_A'(\lambda) = \frac{E(Xe^{\lambda X})}{AE(e^{\lambda X})} = 1.$$

The two difficulties are easily taken care of. (a). The $X_j$ have a common distribution. To make $X_j$ independent corresponds to dividing the original interval $(0,N)$ into intervals of length $A$ and suppose no connection between different intervals. As is easily seen, this gives the same limit as $N \to \infty$ and then $A \to \infty$. (b). $X(A)$ depends on $N$ but the limit $\psi_A(\lambda)$ exists uniformly in $\lambda$ over finite range which is sufficient for the above computation. Therefore, $X(A)$ is

$$X(A) = \frac{1}{2\pi}\int_0^{2\pi A} \hat{\mu}(F(x;\xi))dx,$$

where

$$F(x,\xi) = \lim_{N\to\infty} \frac{1}{\sqrt{N}} \sum_{\nu=0}^{\infty} \hat{\Phi}\left(\frac{\nu}{N}\right)^{\frac{1}{2}}\left(\xi_\nu \cos \frac{\nu x}{N} + \xi_{-\nu} \sin \frac{\nu x}{N}\right).$$

Lemma. $F(x;\xi) \in C^\infty$ with probability 1, and $F(x;\xi)$ and $F(y;\xi)$ are independent if $|x-y| > 2\pi$ when $\Phi$ has support in $(-\pi,\pi)$.

Proof. $F(x;\xi)$ is a stochastic integral and the first statement is a well-known fact. The $F(x;\xi)$ are clearly normal and

$$E(F(x;\xi)F(y;\xi)) = \Phi(x-y).$$

It remains to study (17) as $A \to \infty$.

We divide $(0,2\pi A)$ into intervals $u_1, v_1, u_2, v_2, \ldots$ of equal length $>2\pi$. Let $\psi_j(t)$ be some sequence of functions dense in $C_0^\infty(u)$. We describe $F(x;\xi)$ by its moments,

$$\alpha_k(u) = \int_u F(t,\xi)\psi_k(t)dt.$$

Given $\alpha_1,\ldots,\alpha_s$ for $u = u_1$ and $u = u_2$. Let

$$K_s(\alpha(u_1),\alpha(u_2)) = E_\alpha e^{\frac{1}{2}\lambda\int_{u_1}\hat{\mu}(F)dt} \cdot e^{\frac{\lambda}{2\pi}\int_{v_1}\hat{\mu}(F)dt} \cdot e^{\frac{1}{2}\lambda\int_{u_2}\hat{\mu}(F)dt}$$

where $E_\alpha$ is expectation under the conditions $\alpha(u_1),\alpha(u_2)$. We define the operator,

$$T_s f = \int_{u_1} K_s(\alpha^1,\alpha^2)f(\alpha^1)dP(\alpha^1).$$

$T_s f$ is clearly a bounded operator from $L^2(u_1,dP)$ to $L^2(u_2,dP)$. Furthermore, when $s \to \infty$, $T_s$ converges to a completely continuous operator. This follows immediately from $K_s$ being uniformly bounded and the lemma.

$T$ is not symmetric. However, there exists an involution, $\alpha \to \alpha^*$, which consists in changing the order on the x-axis. This preserves measures and $K(\alpha,\beta) = K(\alpha^*,\beta^*)$. We form

$$K_0(\alpha,\beta) = \frac{1}{4}[K(\alpha,\beta) + K(\alpha^*,\beta) + K(\alpha,\beta^*) + K(\alpha^*,\beta^*)].$$

The corresponding operator $T_o$ has an isolated largest eigenvalue by the previous lemma. As before, we have

$$E(e^{\lambda X(A)}) \sim \lambda^A;$$

thus $\log\lambda$ is essentially the free energy $F(\beta)$ in the theorem. The operator $T_0$ depends analytically on $\beta$ and so the result follows.

ON SPECTRAL SYNTHESIS IN $\mathbb{R}^n$, $n \geq 2$

Yngve Domar
Uppsala University

1. This introduction surveys what is known about spectral synthesis for smooth sets in $\mathbb{R}^n$.

Let $B$ be a Banach space of continuous, complex-valued functions on $\mathbb{R}^n$. We assume that the Schwartz space $\mathcal{D}(\mathbb{R}^n)$ is a dense subspace of $B$, and that convergence in $B$ implies pointwise convergence. For every closed set $E \subseteq \mathbb{R}^n$ we introduce the following three subspaces of $B$:

$B_1(E)$, the (closed) subspace of all $f \in B$ which vanish on $E$,

$B_2(E)$, the closure in $B$ of the space of all $f \in \mathcal{D}(\mathbb{R}^n)$ which vanish on $E$,

$B_3(E)$, the closure in $B$ of the space of all $f \in \mathcal{D}(\mathbb{R}^n)$, each of which vanishes in some neighborhood of $E$.

Then $B_1(E) \supseteq B_2(E) \supseteq B_3(E)$. Following the usual definition (see for instance [1]), we say that $E$ is of synthesis with respect to $B$, if $B_1(E) = B_3(E)$. As was pointed out by C. S. Herz [16], it is also very natural to study the following weaker property: we say that $E$ is of weak synthesis with respect to $B$, if $B_1(E) = B_2(E)$. Thus synthesis implies weak synthesis. Whether the converse holds depends on the choice of the space $B$.

We shall only discuss the case that $B$ is the space $A(\mathbb{R}^n)$ of Fourier transforms of functions in $L^1(\mathbb{R}^n)$, with the norm inherited from that space. Evidently all postulates for $B$ are then fulfilled. We have the following well-known result:

For $A(\mathbb{R}^n)$ synthesis and weak synthesis are equivalent properties in each of the following cases:

1°. when $E = \overline{E^o}$ (the closure of the interior of $E$),

2°. when $n = 1$ or $2$,

3°. when $E$ is a subset of an arc of a rectifiable curve.

The proof of 1° follows from the fact that, if $E = \overline{\mathring{E}}$, $f \in B_2(E)$ implies that all derivatives of $f$ vanish on E. Then $f$ can be approximated in $\mathcal{D}(\mathbb{R}^n)$ by a sequence of functions in $B_3(E)$. Since the topology in $\mathcal{D}(\mathbb{R}^n)$ is stronger than the topology in $A(\mathbb{R}^n)$, this gives 1°. The remaining statements follow from a well-known technique, developed by A. Beurling and H. Pollard (cf., Herz [15]).

The first counterexample to spectral synthesis was given in 1948 by L. Schwartz in [24], where he proves that the sphere $S^{n-1} \subseteq \mathbb{R}^n$ is a set of non-synthesis, if $n \geq 3$. His proof, however, does not reveal whether $S^{n-1}$ is of weak synthesis or not. This question was answered later, by Herz and N. Varopoulos. In 1958 Herz proved that $S^1 \subseteq \mathbb{R}^2$ is a set of synthesis [15], and in 1966 Varopoulos extended the method of Herz to higher dimensions, obtaining as a partial result that $S^{n-1} \subseteq \mathbb{R}^n$ is of weak synthesis, if $n \geq 3$ [25]. Thus the notions of synthesis and weak synthesis do not coincide, if $n \geq 3$.

It is natural to ask if there are any sets at all which are not of weak synthesis. The answer is "yes" for every $\mathbb{R}^n$. A set of non-synthesis on $\mathbb{R}$ was constructed by Malliavin [21] in 1959, and, by imbedding $\mathbb{R}$ in $\mathbb{R}^n$, his set gives a set of non-synthesis in $\mathbb{R}^n$, for every $n$. By 2° and 3° it is not even of weak synthesis. Other constructions of sets of non-synthesis on $\mathbb{R}$ have been made later by Varopoulos [26] and T. Körner [19] (cf., R. Kaufman [17]).
The counter-examples in $\mathbb{R}^n$, which we obtain in this way, are all complicated sets, with no obvious properties of structure and regularity.

It is thus a natural thing to ask whether all sufficiently smooth closed subsets of $\mathbb{R}^n$ are of weak synthesis. $S^{n-1} \subseteq \mathbb{R}^n$ is in agreement with this conjecture, and the following known result seems to favor it: $E \subseteq \mathbb{R}^n$ is of synthesis if $E = \overline{\mathring{E}}$ and if the boundary of $E$ is an $(n-1)$-dimensional $C^1$ manifold.

Here is a sketch of the proof. A partition of unity shows that it is enough to prove that every point in the boundary has a neighborhood such that functions in $B_1(E)$ with support in this neighborhood can be approximated by functions in $B_3(E)$. Choosing the neighborhood small, this can be accomplished by first approximating the function with suitable translates.

In view of this result it is natural to turn our attention to the case when $E^{\circ}$ is empty. From now on we shall assume that $E$ is a smooth manifold in $\mathbb{R}^n$, of dimension $\leq n-1$, or a subset of such a manifold such that it is (in the restriction topology of the manifold) the closure of its interior and has a $C^1$ boundary.

For such manifolds we already have one special result: $S^{n-1} \subseteq \mathbb{R}^n$ is of weak synthesis. But the method of Herz and Varopoulos cannot be generalized directly to general manifolds. The reason is that they rely on the facts that $f \in A(\mathbb{R}^n)$ implies $f \circ \varphi \in A(\mathbb{R}^n)$, with unchanged norm, if $\varphi$ is a nondegenerate affine mapping, and that $S^{n-1}$ is an orbit of a continuous group of affine mappings, the group of rotations around the origin. This is how their basic idea implies the weak synthesis property. A bounded linear functional on $A(\mathbb{R}^n)$, annihilating $B_2(S^{n-1})$, can be regarded as an element in $\mathcal{D}'(\mathbb{R}^n)$ with support in $S^{n-1}$. Using averages of rotations of $\nu$, it is possible to construct a sequence of bounded Borel measures, supported by $S^{n-1}$, converging in the weak* sense to $\nu$. The measures annihilate $B_1(S^{n-1})$, and thus $\nu$ does the same. The same argument can be used for other manifolds which are orbits of continuous groups of nondegenerate affine mappings in $\mathbb{R}^n$, but this family of manifolds is of course very small (cf., F. Lust [20]).

When attempting to generalize to other manifolds, the great obstacle is that the nondegenerate affine mappings are the only mappings which leave $A(\mathbb{R}^n)$ invariant (A. Beurling and H. Helson [3]).

The basic idea of Herz can nevertheless be exploited. If $E$ can be regarded as the orbit of a group of mappings of $\mathbb{R}^n$, then, even if the individual mappings are not operators in $A(\mathbb{R}^n)$, there is still the possibility that certain averages are operators converging to the identity operator. The adjoints of the average operators give in this case a weak* convergent sequence, which can be used to prove weak synthesis exactly in the same way as in the discussion on $S^{n-1}$ above.

This programme may sound easy, but it is in general difficult to carry out. First, it is necessary to find a good group of mappings, and then the essential difficulty is to prove uniform boundedness of the operators given by the averages. Even for smooth curves in $\mathbb{R}^2$ problems arise. The method fails if there is a point of the curve where the curvature changes its sign — then no satisfactory group of mappings can be found. Even if the curvature does not vanish at all on the curve, the mappings must be chosen with care: the orbits have to form a family of curves such that the isoclines (the loci for points with the same tangent direction) are line segments. With the proper choice of mappings and careful estimation of the operator norm, the task can be carried out. In [6] it was proved, using mappings with parallel curves as orbits, that a simple $C^2$ curve in $\mathbb{R}^2$ is of synthesis if its curvature does not vanish. In [7], basically the same method gave that a $C^\infty$ manifold with Gaussian curvature $\neq 0$, or a certain subset thereof, is of weak synthesis; this is an extension of the result of Varopoulos. In the following sections of this paper we shall show that this general idea can be applied in $\mathbb{R}^2$ to the study of general smooth families of curves with curvature $\neq 0$ and with the isoclines being line segments. This will give us interesting relations between the spectral properties of sets in $\mathbb{R}^2$ and $\mathbb{R}$.

Let us return now to the result in [6] for curves in $\mathbb{R}^2$. It is possible to use standard results in the theory of spectral synthesis to find the following extension: let $E \subseteq \mathbb{R}^2$ be a set such that the removal of a certain denumerable point set splits $E$ into disjoint pieces which are either line segments or simple $C^2$ curves with curvature $\neq 0$; then $E$ is of synthesis. In view of this, it is natural to ask whether the curvature assumption is needed at all. Is every simple $C^\infty$ curve of synthesis? This would then agree with one general conjecture that all sufficiently smooth sets are of weak synthesis.

However, it is possible to find a $C^\infty$ function from $\mathbb{R}$ to $\mathbb{R}$, the graph of which is a set of non-synthesis in $\mathbb{R}^2$ (thus at the same time a set not of weak synthesis). This was done in [12], starting from a counterexample on $\mathbb{R} \subseteq \mathbb{R}^2$ and extending the corresponding function in $A(\mathbb{R})$ to $A(\mathbb{R}^2)$ so that it vanishes on a set of the desired type.

This example shows that our conjecture, that $C^\infty$ continuity yields weak synthesis, is not true. The curvature assumptions which were needed for our method to work seem to be in the nature of things. One has to restrict the conjecture, say, to analytic manifolds in order to have hope for an affirmative answer — maybe there is, in that case, a different approach which avoids curvature assumptions.

The non-synthesis curve in $\mathbb{R}^2$ is of some interest, too, in the respect that it can be used to disprove certain conjectures for the notion of synthesis in $\mathbb{R}^2$ (thus also for weak synthesis). Thus we find, using $1^\circ$,

A. A set of synthesis in $\mathbb{R}^2$ may have a boundary of non-synthesis.

B. The intersection of two sets of synthesis in $\mathbb{R}^2$ may be of non-synthesis.

C. A $C^\infty$ map of a set of synthesis in $\mathbb{R}^2$ may be of non-synthesis.

The special problems which arise, when one attempts to treat manifolds of dimension $\leq n-2$ in $\mathbb{R}^n$, $n \geq 3$, are already apparent in the discussion of curves in $\mathbb{R}^3$. Mappings, the orbits of which are parallel curves, are natural to try also in this case; but they do not give average operators which are bounded operators on $A(\mathbb{R}^n)$, as was proved by R. Gustavsson [14]. Only a certain modification of the approach led to the desired result: a simple $C^3$ curve in $\mathbb{R}^3$, with torsion $\neq 0$, is of synthesis [10].

It is believed, however, that the original method should work unchanged with a more subtle choice of mappings, giving a different family of orbits. One should look for a set-up where the orbits have the property that loci for points with the same binormal are line segments while loci for points with the tangent parallel to any fixed plane should be planes. Thus we are led to a differential geometric problem: can every sufficient smooth curve in $\mathbb{R}^3$ with torsion $\neq 0$ be incorporated in a family of curves of the mentioned type? Similar conditions appear in still higher dimensions. It should be stressed that a solution of the differential geometric construction problems does not guarantee extensions of our results. Estimates of the operator norms have to be done, but after a positive reply of the geometric question there is hope that the method would work.

Thus we are left with problems in $\mathbb{R}^n$ where differential geometric concepts are involved. This seems to be a general feature in the study of $A(\mathbb{R}^n)$ and related spaces in their relation to smooth manifolds, as can also be seen from the results in [8], [9] and [11].

2. Let us form the mapping

$$[0,1] \times [-2,2] \ni (\sigma,t) \to (x(\sigma,t),\ y(\sigma,t)) \in \mathbb{R}^2 ,$$

where

(1) the mapping is injective and in $C^\infty$

(2) $$x_\sigma y_t \neq x_t y_\sigma$$

(3) $$x_t y_{tt} \neq x_{tt} y_t$$

(4) $x(\sigma,t)$ and $y(\sigma,t)$ are of the form

$$x(\sigma,t) = \sigma g(t) + j(t)$$

$$y(\sigma,t) = \sigma k(t) + \ell(t),$$

where

$$g'(t)\ell'(t) = k'(t)j'(t) .$$

We shall make some remarks on the properties of such a mapping. Properties (1) and (2) are regularity conditions, implying that there is an inverse in $C^\infty$ and also implying that the two curve families, obtained as maps of the line segments $\sigma$ = constant and $t$ = constant, are smooth $C^\infty$ families with non-zero intersection angles. Condition (3) says that the curves in the first family (the images of $\sigma$ = constant) have curvature $\neq 0$. Condition (4) implies that the tangent direction of these curves depends only on $t$. Thus the isoclines are line segments

(5) $$\begin{cases} x = \sigma g(t_o) + j(t_o) \\ y = \sigma k(t_o) + \ell(t_o) , \end{cases}$$

$\sigma \in [0,1]$, or finite unions of such sets. It is important to observe that the mapping from [0,1] to any isocline, given by (5), is affine.

Conversely, every $C^\infty$ curve family, with curvatures $\neq 0$ and with all isoclines the unions of line segments, can be locally identified with the images of the segments $\sigma$ = constant, for some mapping of this form. The reason is that the corresponding differential equation, a "generalized Clairaut equation," can be integrated explicitly, giving solutions of the desired form. In view of what was remarked in Section 1, it is thus natural to make a thorough investigation of our mapping in relation to $A(\mathbb{R}^2)$.

Let us first introduce some notation. For any closed set $F \subseteq \mathbb{R}^n$, $A(F) = A(\mathbb{R}^n)/B_1(F)$. Thus it is the Banach space of restrictions to $F$ of $A(\mathbb{R}^n)$. Let $E$ be the image of the whole rectangle $[0,1] \times [-2,2]$ for our mapping, and let $E'$ be the image of $[0,1] \times [-1,1]$. For every $s \in [-1,1]$, $T_s$ is the mapping of $E'$ into $E$ which sends $(x(\sigma,t), y(\sigma,t))$ into $(x(\sigma,t-s), y(\sigma,t-s))$. We may call $T_s$ "translation by $s$ along the curves."

If $f \in A(E)$, it does not follow in general that $f \circ T_s \in A(E')$. This is seen from the theorem of Beurling and Helson [3], which P. Brenner [4] has shown to hold in a local version, applicable to $A(E')$. His result says that the implication holds only if $T_s$ is the restriction of an affine mapping of $\mathbb{R}^2$.

We shall go one step further, introducing a concept which can be described as "convolution along the curves." For $f \in A(E)$, $\varphi \in \mathcal{D}([-1,1])$, we form

$$T_\varphi f = \int_{-1}^{1} f \circ T_s \varphi(s) ds .$$

Thus $T_\varphi f$ is the function on $E'$ taking the value

$$\int_{-1}^{1} f(x(\sigma,t-s), y(\sigma,t-s)) \varphi(s) ds$$

at the point $(x(\sigma,t), y(\sigma,t))$.

It turns out that $T_\varphi f \in A(E')$, and we shall in fact prove a more general result. Let us denote by $\varphi_h$, for $h \in ]0,1]$, the function on $\mathbb{R}$ such that $\varphi_h(t) = (1/h)\varphi(t/h)$. Then the following holds:

<u>Theorem 1</u>. Under our assumptions, we have

$$T_\varphi f \in A(E').$$

If $\int_{\mathbb{R}} \varphi(t) dt = 1$, then

$$T_{\varphi_h} f \to f\big|_{E'},$$

in $A(E')$, as $h \to 0$.

This is a general theorem that can be used in discussions of spectral synthesis problems on $\mathbb{R}^2$, as sketched in Section 1. In Section 4 we shall see how Theorem 1 can be exploited in this direction.

The proof of Theorem 1 is given in Section 3. Here we shall just give some indication on the ideas in it, and prove a simple lemma.

In the proof we use the following elementary facts and techniques:

1) If $\chi$ is a bounded continuous character on $\mathbb{R}^2$ and $F \subseteq \mathbb{R}^2$, then $f \in A(F)$ implies $\chi f \in A(F)$, and

$$\|\chi f\|_{A(F)} = \|f\|_{A(F)}.$$

(In fact, multiplication with $\chi$ corresponds to translation on the dual $\mathbb{R}^2$, and this does not change the norm.)

2) If $\Phi$ is a bijective affine map of $F \subseteq \mathbb{R}^2$ and $f \in A(F)$, then $f \circ \Phi^{-1} \in A(\Phi(F))$ and

$$\|f \circ \Phi^{-1}\|_{A(\Phi(F))} = \|f\|_{A(F)}.$$

(Follows from the invariance of $A(\mathbb{R}^2)$ and its norm under non-degenerate affine mappings.)

3) Let $F$ be a nice and not too large compact subset of $\mathbb{R}^2$; then bounds of $f \in C^2(F)$ together with its derivatives of first and second order give a bound for $\|f\|_{A(F)}$ which is independent of $F$.

(This is made more precise in Lemma 1, which follows.)

4) Let $(F_n)_1^\infty$ be compact subsets of $\mathbb{R}^2$. Put $F = \cup F_n$, and suppose $F$ is closed. Under certain conditions, implying among other things that there is a substantial overlapping between the sets, there

is a constant $C$, such that if $f$ is defined on $F$ and $f|_{F_n} \in A(F_n)$, then $f \in A(F)$ and

$$\|f\|_{A(F)} \le C \cdot \sum \|f|_{F_n}\|_{A(F_n)}.$$

(The needed details are given in the beginning of the proof in Section 3.).

As a preparation for Lemma 1, we shall prove an inequality, due to F. Carlson [5] (who proved the corresponding one-dimensional inequality), Beurling [3] (who found a new proof, extendable to higher dimensions), and B. Kjellberg [18] (who made such extensions):

Inequality: Let $f, f_{xx}, f_{yy} \in L^2(\mathbb{R}^2)$, where the derivatives are defined in distributional sense. Then $f$ can be altered on a set of measure 0, so that $f \in A(\mathbb{R}^2)$ and

$$(6) \qquad \|f\|_{A(\mathbb{R}^2)} \le C\left[\|f\|_{L^2(\mathbb{R}^2)} \cdot (\|f_{xx}\|_{L^2(\mathbb{R}^2)} + \|f_{yy}\|_{L^2(\mathbb{R}^2)})\right]^{1/2},$$

where $C$ is an absolute constant.

Sketch of the proof: By Schwarz' inequality we have, for $\lambda > 0$,

$$\int_{\mathbb{R}^2} |\hat{f}|\,d\xi\,d\eta = \int_{\mathbb{R}^2} (\lambda + \xi^4 + \eta^4)^{1/2}|\hat{f}| \cdot (\lambda + \xi^4 + \eta^4)^{-1/2} d\xi\,d\eta \le$$

$$\le \left(\int_{\mathbb{R}^2} (\lambda + \xi^4 + \eta^4)|\hat{f}|^2 d\xi\,d\eta\right)^{1/2} \cdot C_0 \cdot \lambda^{-1/4} =$$

$$= C_0\left(\lambda^{1/2} \cdot \int_{\mathbb{R}^2} |\hat{f}|^2 d\xi\,d\eta + \lambda^{-1/2} \cdot \int_{\mathbb{R}^2} |\hat{f}|^2(\xi^4 + \eta^4)d\xi\,d\eta\right)^{1/2},$$

where $C_0$ is an absolute constant. Choosing $\lambda$ so that the last member is minimized and then using Parseval's relation, the desired inequality is obtained.

Lemma 1. Let $C \ge 1$, and let $F$ be a compact subset of $\mathbb{R}^2$, satisfying $F = \overline{F^{\circ}}$ and $\operatorname{diam}(F) \leq C$. Suppose furthermore that every pair $z_1, z_2$ of points in $F$ can be joined by a curve in $F$ with length $\leq C|z_1 - z_2|$. Then $f \in C^2(F)$ implies $f \in A(F)$, and there exists a constant $K$, depending only on $C$, such that

$$\|f\|_{A(F)} \leq K\|f\|_{C^2(F)},$$

where

$$\|f\|_{C^2(F)} = \sum_{|\alpha| \leq 2} \sup_F |f^{\alpha}|.$$

<u>Proof</u>: $F$ is included in a circular disc of radius $C$. Since $C \geq 1$, it is obvious that there exists a function $\varphi \in C^2(\mathbb{R}^2)$, with the value 1 on $F$, vanishing outside a circle with radius $2C$, and satisfying

$$\|\varphi\|_{C^2(\mathbb{R}^2)} \leq K_1, \tag{7}$$

where $K_1$ is an absolute constant.

By a quantitative version of Whitney's extension theorem [27], there is a constant $K_2$, depending on $C$, such that $f$ can be extended to $\mathbb{R}^2$ and

$$\|f\|_{C^2(\mathbb{R}^2)} \leq K_2 \cdot \|f\|_{C^2(F)}. \tag{8}$$

Then $f\varphi$ is also an extension of $f$. It vanishes outside a circular disc of radius $2C$, and in the disc it satisfies, by (7) and (8),

$$\|f\varphi\|_{C^2(\mathbb{R}^2)} \leq K_3\|f\|_{C^2(F)},$$

where $K_3$ is a constant depending only on $C$. An application of the inequality (6) to $f\varphi$ gives

$$\|f\varphi\|_{A(\mathbb{R}^2)} \leq K\|f\|_{C^2(F)},$$

where $K$ depends only on $C$, and the lemma is proved.

It should be mentioned that in the applications which we shall make of the lemma, Whitney's theorem is dispensable since in these cases it is easy to make explicit extensions of the considered functions.

3. We shall now prove Theorem 1.

Let us first point out that it suffices to show that $T_\varphi f \in A(E')$ and that there exists a constant $C$ such that

$$\|T_{\varphi_h} f\|_{A(E')} \leq C\|f\|_{A(E)} , \tag{9}$$

independently of $f$ and $h$.

The reason for this is that $\mathcal{D}(\mathbb{R}^2)$ is a dense subspace of $A(\mathbb{R}^2)$. Hence $C^\infty(E)$ is dense in $A(E)$. For every $f \in C^\infty(E)$, it is easy to see that the condition,

$$\int_{\mathbb{R}} \varphi(t)dt = 1,$$

implies that $T_{\varphi_h} f - f|_{E'}$ and its derivatives of all orders converge uniformly to $0$ on $E'$. Thus, by the lemma,

$$T_{\varphi_h} f \to f|_{E'} , \quad \text{in } A(E),$$

as $h \to 0$. By the density, the same result holds for every $f \in A(E)$.

Furthermore it suffices to prove (9) when $f$ is a (bounded continuous) character $\chi$. In fact, in that case, if

$$f(x,y) = \int_{\mathbb{R}^2} e^{-i(x\xi + y\eta)}\hat{f}(\xi,\eta)d\xi\, d\eta$$

on $E$, with $\hat{f} \in L^1(\mathbb{R}^2)$, then

$$T_{\varphi_h} f = \int_{\mathbb{R}^2} T_{\varphi_h}(e^{-i(x\xi + y\eta)})\hat{f}(\xi,\eta)d\xi\, d\eta$$

on $E'$; and hence

$$\|T_{\varphi_h} f\|_{A(E')} \le C \int_{\mathbb{R}^2} |\hat{f}(\xi,\eta)|\, d\xi\, d\eta \ .$$

Varying $f$, we obtain (9).

Thus we must prove that

$$\|T_{\varphi_h} \chi\|_{A(E')}$$

is uniformly bounded, as $\chi$ varies through the set of characters on $\mathbb{R}^2$, and $h$ varies in $[-1,1]$.

We shall restrict ourselves to a one-parameter set of characters, and, moreover, shall assume that certain special conditions hold for the mapping. Having proved uniform boundedness for this particular case, we can then obtain it in the general situation simply by reviewing the proof and seeing how the uniform bound depends on the various data involved.

We assume that the set of characters under discussion is $\{\chi_a : (x,y) \to e^{iay}$ where $a > 0\}$, and that the mapping fulfills the conditions stated in the beginning of this section as well as the following conditions:

(10) $$k(0) = 1, \quad g(0) = j(0) = \ell(0) = 0,$$

(11) $$x_t(\sigma,t) > 0, \quad \text{for every } (\sigma,t)$$

(12) $$y_t(\sigma,-t) < 0 < y_t(\sigma,t), \quad \text{if} \quad \sigma \in [0,1], \quad t \in \,]0,2] \ .$$

(13) $$y_{tt}(\sigma,0) > 0 \ .$$

We want to find an upper bound of the $A(E')$-norm of the function $T_{\varphi_h} \chi_a$, the value of which at $(x(\sigma,t),\ y(\sigma,t))$ is

(14) $$\int_{\mathbb{R}} e^{iay(\sigma,t-s)} \frac{1}{h} \varphi\left(\frac{s}{h}\right) ds \ .$$

The bound has to be uniform in $a$ and $h$.

We cannot apply the lemma directly, for $T_{\varphi_h}\chi_a$ behaves differently in different parts of $E'$, and for that reason we have to split up the set.

Let $N$ be an integer $\geq 0$. We put $I_o = [-2^{-N}, 2^{-N}]$, and, for $0 < n \leq N$,

$$I_n = [2^{-n-1}, 2^{-n+1}], \quad I_{-n} = -I_n .$$

Then $\cup I_n$ forms a covering of $[-1,1]$. Let $E_n$ be the map of $[0,1] \times I_n$. Thus $\cup E_n$ forms a covering of $E'$ by means of overlapping sectors.

We shall now make a partition of unity: for every $n$, $-N \leq n \leq N$, we construct a function $\varphi_n \in C^2(\mathbb{R}^2)$, which vanishes on $E' \setminus E_n$, such that $\sum_{-N}^{N} \varphi_n = 1$ on $E'$. This can be done in such a way that, for any $n$, there exists an affine mapping which takes $\varphi_n$ into a function with $C^2$-norm $\leq C$ and support having area $\leq C$, where $C$ is a constant, <u>independent of $n$</u>. The proof of this is left to the reader.

By the lemma in Section 2, this means that the $A(\mathbb{R}^2)$-norms of the functions $\varphi_n$ are uniformly bounded by some constant $D$. Hence, since

$$\|\varphi\psi\|_{A(\mathbb{R}^2)} \leq \|\varphi\|_{A(\mathbb{R}^2)} \|\psi\|_{A(\mathbb{R}^2)}, \quad \varphi, \ \psi \in A(\mathbb{R}^2),$$

we obtain, for every $g \in A(E')$,

$$\|g\|_{A(E')} = \|\sum g\varphi_n\|_{A(E')} \leq \sum \|g\varphi_n\|_{A(E')} \leq$$

$$\leq \sum \|g\|_{A(E_n)} \|\varphi_n\|_{A(\mathbb{R}^2)} \leq D \sum \|g\|_{A(E_n)} .$$

Applied to $T_{\varphi_h}\chi_a$, this gives

(15) $$\|T_{\varphi_h}\chi_a\|_{A(E')} \le D \sum_{-N}^{N} \|T_{\varphi_h}\chi_a\|_{A(E_n)} .$$

Thus it suffices to show that, for every $a$ and $h$, $N$ can be chosen in such a way that the right hand member of (15) is bounded, uniformly in $a$ and $h$.

We shall choose $N$ as the largest integer $\geq 0$, such that

$$2^{-N} \geq \mathrm{Max}(4h, \frac{1}{ah}) .$$

If no such integer exists, we take $N = 0$.

The estimate of the right hand member of (15) is organized as follows: first we estimate the term with index $0$, in the case that $\frac{1}{ah} \geq 4h$, then we take the $n$-th term (in both cases), and finally the term $n = 0$, when $\frac{1}{ah} < 4h$.

I. Let $\frac{1}{ah} \geq 4h$, and put $b = 2^{-N}$. By property (2) in Section 2, we can, without change of norm, multiply the function (14) with $\overline{\chi}_a$, giving

$$\int_{\mathbb{R}} e^{ia(y(\sigma,t-s) - y(\sigma,t))} \frac{1}{h}\varphi(\frac{s}{h})\, ds =$$

$$= \int_{\mathbb{R}} e^{ia(y(\sigma,t-hs) - y(\sigma,t))}\varphi(s)ds ;$$

and we have to estimate its norm in $A(E_o)$.

Changing the $x$-variable, affinely, and at the same time changing the $t$-parameter, we see that this is the same problem as estimating the $A(F)$-norm of the function with value

$$\int e^{ia(y(\sigma,bt-hs) -y (\sigma,bt)}\varphi(s)ds,$$

where

(16) $$\begin{cases} x = \frac{1}{b}\, x(\sigma,bt) \\ y = y(\sigma,bt) \end{cases}$$

and $F$ is the image of $[0,1] \times [-1,1]$ under this mapping.

Due to the conditions (10) - (13) the mapping (16) is in $C^\infty$, with each individual derivative uniformly bounded in $\sigma, t$ and $b$. Furthermore its functional determinant is in similar fashion bounded away from 0. Hence the inverse is bounded in $C^\infty$.

This means that the region $F$ in the xy-plane where the A-norm shall be taken satisfies the condition of the lemma in Section 2 in a uniform way. Thus the $A(F)$-norm is uniformly bounded if the $C^2$-norm is uniformly bounded, which in turn is equivalent to the $C^2$-norm of the function, considered as a function of $(\sigma,t)$, being uniformly bounded in $[0,1] \times [-1,1]$. Since

$$b = 2^{-N} < \frac{2}{ah},$$

it suffices to show that the function

$$\psi(\sigma,t,s) = \frac{1}{hb}(y(\sigma,bt-hs) - y(\sigma,bt)),$$

considered as a function of $(\sigma,t)$, is uniformly bounded in $C^2$, uniformity now considered with respect to $s$, $b$ and $h$.

This is easy to prove, using the explicit expression for $y(\sigma,bt)$ and Lagrange's mean value theorem. For instance, the boundedness of the function follows from

$$\psi(\sigma,t,s) = \frac{1}{hb}\cdot(-hs)\cdot y_t(\sigma,bt-\theta hs),$$

where $0 < \theta < 1$, and since $|s| \leq 1$, $h < b$, condition (12) gives the boundedness. The remaining verifications are left to the reader.

II. We proceed as in I, the difference being that the affine mapping of $E_n$ is made in a direction orthogonal to the isocline corresponding to $t = 2^{-n}$, while keeping this line segment fixed. As before everything proceeds in the same way up to the proof that

$$\int e^{iahb\psi(\sigma,t,s)}\varphi(s)ds$$

is uniformly bounded in $C^2$, considered as function of $(\sigma,s)$ in $[0,1]\times[-1,1]$. This time $b(=2^{-n})$ is not comparable with $1/(ah)$, and we have to proceed in a different way.

A further discussion of $\psi(\sigma,t,s)$ shows that it is in $C^3$ uniformly, considered as function of $(\sigma,t,s) \in [0,1]\times[-1,1]^2$. Furthermore $\psi_s$ is uniformly bounded away from $0$ due to (12) and (13) and the assumption (3) on the curvature of the curves. Hence, $\psi(\sigma,t,s)$ can be introduced as a new variable $u$ of integration, giving

$$\int e^{iahb\xi}\Phi(\sigma,t,\xi)d\xi ,$$

where $\Phi \in C^3$ in a uniform way and vanishes outside a uniformly bounded interval of the $\xi$-axis.

Standard estimates (with partial integration) show that this function of $(\sigma,t) \in [0,1]\times[-1,1]$ has its $C^2$-norm bounded by

$$\leq \frac{C}{ahb} \leq C2^{n-N} ,$$

where $C$ is a uniform constant. This shows that

$$\sum_{-N}^{-1} + \sum_{1}^{N} \|T_{\varphi_h} \chi_a\|_{A(E_n)} \leq 4C.$$

III. Thus it remains only to discuss $E_o$, when $4h > \frac{1}{ah}$. It is convenient to change the parameter $t$ to a new parameter $\xi$ so that

$$y(\sigma,t) = \sigma + \xi^2 .$$

This is done simply by observing that

$$y(\sigma,t) = \sigma + t^2A(\sigma,t) ,$$

where $A(\sigma,t)$ is a positive function in $C^\infty$ (due to our assumptions (1) - (4), (10) - (13)), and putting $\xi = t\sqrt{A(\sigma,t)}$. An easy computation shows that the Jacobian does not vanish and thus that

$$t = \xi B(\sigma,\xi) ,$$

where $B$ is in $C^\infty$.

Hence we obtain

$$\int_{\mathbb{R}} e^{iay(\sigma,t-s)} \frac{1}{h}\varphi(\frac{s}{h})ds =$$

$$= \int_{\mathbb{R}} e^{iay(\sigma,s)} \frac{1}{h}\varphi(\frac{t-s}{h})ds =$$

$$= \int_{\mathbb{R}} e^{ia(\sigma+\xi^2)} \frac{1}{h}\varphi(\frac{t-\xi B(\sigma,\xi)}{h})C(\sigma,\xi)d\xi =$$

$$= e^{ia\sigma}\cdot \int_{\mathbb{R}} e^{iah^2\xi^2} \varphi(\frac{t}{h} - \xi B(\sigma,\xi))C(\sigma,\xi)d\xi ,$$

where $C$ is in $C^\infty$.

By the submultiplicativity of the norm in $B(E_o)$, we can consider each factor at a time. Let us start with the second factor. Put $2^{-N} = b$, which is between $4h$ and $8h$. Using the same arguments as in I and II we find that it suffices to have a bound in $C^2$ for the function

$$(\sigma,t) \mapsto \int_{\mathbb{R}} e^{iah^2\xi^2} \varphi(\frac{b}{h}t - \xi B(\sigma,\xi))C(\sigma,\xi)d\xi ,$$

$(\sigma,t) \in [0,1] \times [-1,1]$. The function can be written

$$(17) \qquad (\sigma,t) \mapsto \int_{\mathbb{R}} e^{iah^2\xi^2} D(t,\xi,\sigma)d\xi ,$$

where $D \in C^\infty$, with uniform bound on each derivative and uniform support.

The Fourier transform of

$$\xi \mapsto e^{iah^2\xi^2} ,$$

taken in the distribution sense, is

$$u \mapsto C_o \frac{1}{h\sqrt{a}} e^{-\frac{iu^2}{ah^2}},$$

where $C_o$ is an absolute constant. Via Parserval's relation this implies that the $C^2$-norm of (17) is bounded by

$$C_1 \cdot \frac{1}{h\sqrt{a}},$$

where $C_1$ is a constant, independent of $a$ and $h$.

Thus it suffices to show that there is a uniform constant $C_2$ such that

$$\|e^{ia\sigma}\|_{A(E_o)} \le C_2 h\sqrt{a}\,.$$

Let $m$ be the largest integer $\le h\sqrt{a}$.

We form the intervals

$$J_k = [-b + \tfrac{k}{m} 2b, -b + \tfrac{k+2}{m} 2b],$$

$0 \le k \le m-2$. By the original mapping, the $[0,1]\times J_k$ are mapped into a covering $\cup F_k$ of $E_o$, and it is easy to see that there is a uniform constant $C_3$ such that

$$\|e^{ia\sigma}\|_{A(E_o)} \le C_3 \sum_0^{m-2} \|e^{ia\sigma}\|_{A(F_k)}\,.$$

Hence it suffices to prove that for every interval $J$ of length $\frac{4b}{m}$ on $[-1,1]$, the norm of $e^{ia\sigma}$ in $A(G)$, where $G$ is the map of $J$, is uniformly bounded. Better yet, we prove the same for all intervals of length $\frac{32}{\sqrt{a}}$ (for $\frac{1}{m} \le \frac{1}{h\sqrt{a}}$, $b \le 8h$, and if $a < 32^2$, then $\|e^{ia\sigma}\|_{A(E_o)}$ is uniformly bounded).

This is how $\|e^{ia\sigma}\|_{A(J)}$ can be estimated. The norm is the same as the norm of $e^{ia\sigma}\cdot\chi$, where $\chi$ is any character.

Take any isocline in $J$. The surface $z = \sigma(x,y)$ is developable, meaning that the tangent-planes are the same for all points for which $(x,y) \in J$. This geometric property is very easily proved from the special form of our mapping. Let $z = \ell(x,y)$ be the equation of the tangent plane. Then choose

$$\chi(x,y) = e^{-ia\ell(x,y)},$$

so that we consider

$$\|e^{ia(\sigma-\ell(x,y))}\|_{A(J)} .$$

For simplicity suppose that the isocline mentioned above is the segment [0,1] on the y-axis. We take the affine mapping that fixes y and enlarges x by the factor $\sqrt{a}$. Call the new region J'. By the invariance of the norm we are left to consider

$$\|e^{ia(\sigma(\frac{x}{\sqrt{a}},y) - \ell(\frac{x}{\sqrt{a}},y))}\|_{A(J')},$$

and it is easy to obtain a uniform bound, in the usual way, using the fact that $\ell$ approximates $\sigma$, with error of order 2. The discussion for arbitrary J is similar.

Thus we have proved the uniform boundedness of the norm under the special assumptions stated in the beginning of this section.

Let us now look at these assumptions. One thing they demand is that the curves turn their direction in an angle $<2\pi$. To go from the general case to this situation is easy: one simply represents the original E by a suitable finite union of overlapping subsets, corresponding to a finite union of overlapping subintervals of [-2,2], and solves the problem for each subset. Then we obtain instead a function defined in the image of $[0,1]\times[a,b]$, where [a,b] is a subinterval of [-2,2], and want to prove the smoothing relation in $[0,1]\times[a',b']$, where a' and b' are in the interior of [a,b]. It is very easy to see that the earlier discussion applies to that discussion as well.

One further assumption is that if the character is constant along the tangent to a curve, then the corresponding isocline is perpendicular to the tangent. This is, however, no severe restriction, for the general case can be transferred to that situation by an affine mapping (with determinant bounded away from 0).

Thus, for every $\theta$, we obtain uniform boundedness for the characters

$$(x,y) \mapsto e^{ia(x\cos\theta + y\sin\theta)},$$

$a \in \mathbb{R}^+$ (just choose the coordinate system properly). An inspection of the proof reveals that we have uniform boundedness as well in the parameter $\theta$, and this ends our proof of Theorem 1.

4. We shall give a variant of Theorem 1, valid for certain mappings giving families of <u>closed</u> curves.

Let $\mathbb{T} = 2\pi\,\mathbb{R}/\mathbb{Z}$, and let us take a mapping

$$[0,1]\times\mathbb{T} \ni (\sigma,t) \mapsto (x(\sigma,t),y(\sigma,t)) \in \mathbb{R}^2,$$

satisfying (1), (2), (3), and (4). The image is denoted by $E$. We define, for every $\varphi \in C^\infty(\mathbb{T})$, $T_\varphi$ by

$$T_\varphi f(x(\sigma,t),y(\sigma,t)) = \frac{1}{2\pi}\int_{-\pi}^{\pi} f(x(\sigma,t-s),y(\sigma,t-s))\varphi(s)ds .$$

<u>Theorem 2</u>. Under our assumptions, $f \in A(E)$ implies $T_\varphi f \in A(E)$. Let $\operatorname{supp}\varphi \subseteq\, ]-\pi,\pi[$, $\int_{-\pi}^{\pi}\varphi(t)dt = 1$, and, for $0 < h \le 1$, $\varphi_h(t) = \frac{1}{h}\varphi(\frac{t}{h})$, if $|t| < 2\pi h$, $\varphi_h(t) = 0$, if $t$ is not on that arc. Then

$$T_{\varphi_h} f \to f$$

in $A(E)$, as $h \to 0$.

The theorem follows directly from Theorem 1 after a partition of $E$ and a splitting of $\varphi$ into a sum of functions with smaller supports.

In particular, we can take $\varphi_o \equiv 1$ on $\mathbb{T}$, obtaining the result that if $f \in A(E)$ then the "mean-value along the arcs" $T_{\varphi_o} f \in A(E)$. Of course the obtained function depends on the particular choice of parametric representation.

Conversely, it can be proved that a mapping of the form considered in Theorem 1 can be extended locally at each point $(\sigma,t)$ to a mapping where the curves are closed. Thus it is no loss of generality to restrict the discussion to the case where the curves are closed, and we shall do this from now on.

Let us introduce the Banach space $A_{1/2}(\mathbb{R})$ of Fourier transforms of functions $\hat{f}$, such that $\hat{f}(t)(1+|t|)^{1/2} \in L^1(\mathbb{R})$ and with the norm defined by

$$\|f\|_{A_{1/2}(\mathbb{R})} = \int_{\mathbb{R}} |\hat{f}(t)|(1+|t|)^{1/2} dt .$$

In every closed set $F \subseteq \mathbb{R}$, $I_{1/2}(F)$ is the closed subspace of functions in $A_{1/2}(\mathbb{R})$, vanishing on $F$, and we put

$$A_{1/2}(E) = A_{1/2}(\mathbb{R})/I_{1/2}(E) .$$

Let us denote by $h$ the mapping

$$E \ni (x(\sigma,t),y(\sigma,t)) \mapsto \sigma \in [0,1] .$$

We then have the following theorem.

Theorem 3. $g \in A_{1/2}[0,1]$ is equivalent to $g \circ h \in A(E)$.

This theorem was proved in the case of concentric circles by Gatesoupe [13], and a proof in the general situation was given in [9]. The closed graph theorem implies that the norms in the two spaces are equivalent.

Representing $\varphi \in C^\infty(\mathbb{T})$ by its Fourier series

$$\varphi(t) = \sum_{-\infty}^{\infty} c_n e^{int} = \sum_{-\infty}^{\infty} c_n \varphi_n(t) ,$$

we obtain, for every $(\sigma,t) \in [0,1]\times\mathbb{T}$ and $f \in A(E)$,

$$T_\varphi f(x(\sigma,t),y(\sigma,t)) = \frac{1}{2\pi}\int_{-\pi}^{\pi} \varphi(t-s)f(x(\sigma,s),y(\sigma,s))ds$$

$$= \sum_{-\infty}^{\infty} C_n e^{int}\frac{1}{2\pi}\int_{-\pi}^{\pi} \bar{\varphi}_n(s)f(x(\sigma,s),y(\sigma,s))ds$$

$$= \sum_{-\infty}^{\infty} C_n e^{int}\cdot T_{\varphi_o}(\bar{\varphi}_n\cdot f)(x(\sigma,t),y(\sigma,t)) ,$$

where the convergence is absolute since $C_n = O(n^{-p})$ for every $p$.

If $\varphi_n$ is considered as a function on $E$ by $(\sigma,t)\mapsto\varphi_n(t)$, the lemma in Section 2 gives us that its norm in $A(E)$ is $O(n)$. Hence

$$\|\bar{\varphi}_n\cdot f\|_{A(E)} = O(n),$$

by the submultiplicativity of the norm. By Theorem 2,

$$\|T_{\varphi_o}(\bar{\varphi}_n\cdot f)\| = O(n) ,$$

giving

$$\|\varphi_n T_{\varphi_o}(\bar{\varphi}_n\cdot f)\| = O(n^2) .$$

Hence the series converges in norm, which proves the first part of the following theorem.

<u>Theorem 4</u>. In $A(E)$, every $f$ vanishing on a closed set $F$ of the form $h^{-1}(G)$, for some $G$, can be approximated by linear combinations of functions $\alpha\cdot\beta$, where $\alpha, \beta \in A(E)$, $\alpha$ depends only on $\sigma$, $\beta$ depends only on $t$, and $\alpha$ vanishes on $F$.

For every $\varphi \in C^\infty(\mathbb{T})$ and $f \in A(E)$, the restriction of $T_\varphi f$ to an isocline, considered as function of $\sigma$, is in $A_{1/2}([0,1])$.

The second part of the theorem is proved as follows.

$$\|T_{\varphi_o}(\bar{\varphi}_n\cdot f)\|_{A(E)} = O(n),$$

as proved above. $T_{\varphi_o}(\overline{\varphi}_n \cdot f)$ is constant along the curves, and, hence, its values along an isocline, considered as a function of $\sigma$, has $A_{1/2}([0,1])$-norm of the order $O(n)$. Since $\sum|C_n|n < \infty$, $T_\varphi f$, restricted to an isocline, is in $A_{1/2}([0,1])$.

The second part tells us that the convolution along the curves, which smooths the functions on each individual curve to a $C^\infty$ function, also smooths to some extent the function on every isocline. $f \in A(E)$ implies that its restriction to every isocline (as a function of $\sigma$) is in $A([0,1])$, whereas the restriction of $T_\varphi f$ is in $A_{1/2}([0,1])$.

We shall finally prove a result on spectral synthesis. We say that a set $F \in \mathbb{R}$ is of synthesis for $A_{1/2}(\mathbb{R})$, if every $f \in A(\mathbb{R})$ vanishing on $F$ can be approximated by functions in $\mathcal{D}(\mathbb{R})$ vanishing on individual neighborhoods of $F$.

Take $F \subseteq [0,1]$ and form $h^{-1}(F)$.

<u>Theorem 5</u>. $F$ is of synthesis with respect to $A_{1/2}(\mathbb{R})$ if and only if $h^{-1}(F)$ is of synthesis with respect to $A(\mathbb{R}^2)$.

<u>Proof</u>: By the first part of Theorem 4, synthesis $(A(\mathbb{R}^2))$ for $h^{-1}(F)$ is equivalent to the approximation property (see Section 1) holding for functions in $\mathbb{R}^2$ of the form $\alpha(\sigma)\beta(t)$ in $F$, which in its turn is equivalent to the property holding for functions of the form $\alpha(\sigma)$ in $F$. For the subspace of these functions, we can restrict to approximating by functions coinciding with test functions of the form $\alpha(\sigma)$ in $F$, which is seen by applying the operation $T_{\varphi_o}$. Thus, by Theorem 3, the approximation problem in $A(\mathbb{R}^2)$ for $h^{-1}(F)$ is equivalent to the approximation problem in $A_{1/2}(\mathbb{R})$ for $F$.

<u>Remarks</u>. The result in Theorem 5 is somewhat related to a result by H. Reiter [22,23], which for $\mathbb{R}^2$ can be formulated: $E$ is of

synthesis for $A(\mathbb{R})$ if and only if $E \times \mathbb{R}$ is of synthesis for $A(\mathbb{R}^2)$.

Following ideas in [13], one can carry over the problems of this section to $\mathbb{R}^3$, by introducing the two-parameter family of all curves which lie in planes parallel to the xy-plane and for which the orthogonal projections to the xy-plane are curves in the family we have just considered. Theorems giving relations between spectra for sets in $\mathbb{R}^2$ and $\mathbb{R}^3$ seem within reach, and in particular this could provide methods to analyze the synthesis property for developable surfaces in $\mathbb{R}^3$.

REFERENCES


[1] J. J. Benedetto, Spectral synthesis, New York, 1975.

[2] A. Beurling, Sur les intégrales de Fourier absolument convergentes et leur application à une transformation fonctionelle, IX Congrès des Mathématiciens Scandinaves, Helsinki, 1938(1939), 345-366.

[3] A. Beurling and H. Helson, Fourier-Stieltjes transforms with bounded powers, Math. Scand. 1(1953), 120-126.

[4] Ph. Brenner, Power bounded matrices of Fourier-Stieltjes transforms, Math. Scand. 22(1968), 115-129; Corrections, Math. Scand. 30(1972), 150-151.

[5] F. Carlson, Une inégalité, Ark. Mat. Astr. Fys. 25, B1(1934).

[6] Y. Domar, Sur la synthèse harmonique des courbes de $\mathbb{R}^2$, C. R. Acad. Sci. Paris 270(1970), 875-878.

[7] Y. Domar, On the spectral synthesis problem for $(n-1)$-dimensional subsets of $\mathbb{R}^n$, $n \geq 2$, Ark. Mat. 9(1971), 23-37.

[8] Y. Domar, Estimates of $\|e^{itf}\|_{A(\Gamma)}$, when $\Gamma \subseteq \mathbb{R}^n$ and $f$ is a mean-valued function, Israel J. Math. 12(1972), 184-189.

[9] Y. Domar, Harmonic analysis of a class of distributions on $\mathbb{R}^n$, $n \geq 2$. SIAM J. Math. Anal. 3(1972), 230-245.

[10] Y. Domar, On spectral synthesis for curves in $\mathbb{R}^3$, Math. Scand. 39(1976), 282-294.

[11] Y. Domar, On the Banach algebra $A(\Gamma)$ for smooth sets $\Gamma \subset \mathbb{R}^n$, Comment. Math. Helv. 52(1977), 357-371.

[12] Y. Domar, A $C^\infty$ curve of spectral synthesis, Mathematika 24 (1977), 189-192.

[13] M. Gatesoupe, Sur les transformées de Fourier radiales, Bull. Soc. Math. France, Mémoire 28(1971).

[14] R. Gustavsson, On the spectral synthesis problem for curves in $\mathbb{R}^3$, Uppsala University, Department of Mathematics, Report 1974:6.

[15] C. S. Herz, Spectral synthesis for the circle, Ann. Math. 68 (1958), 709-712.

[16] C. S. Herz, The ideal theorem in certain Banach algebras of functions satisfying smoothness conditions, Proc. Conf. Functional Analysis, Irvine, Calif. 1966, London (1967), 222-234.

[17] R. Kaufman, A pseudofunction on a Helson set II, Astérisque 5 (1973), 231-239.

[18] B. Kjellberg, Ein momentenproblem, Ark. Mat. Fys. Astr. 29 A2(1943).

[19] T. Körner, A pseudofunction on a Helson set I, Astérisque 5 (1973), 3-224.

[20] F. Lust, Le problème de la synthèse et de la p-finesse pour certaines orbites de groupes linéaires dans $A_p(\mathbb{R}^n)$, Studia Math. 39(1971), 17-28.

[21] P. Malliavin, Sur l'impossibilité de la synthèse spectrale sur la droite, C. R. Acad. Sci. Paris, 248(1959), 2155-2157.

[22] H. Reiter, Contributions to harmonic analysis, II, Math. Ann. 133(1957), 298-302.

[23] H. Reiter, Contributions to harmonic analysis, III, J. Lond. Math. Soc. 32(1957), 477-483.

[24] L. Schwartz, Sur une propriété de synthèse spectrale dans les groupes non compacts, C. R. Acad. Sci. Paris 227(1948), 424-426.

[25] N. Th. Varopoulos, Spectral synthesis on spheres, Proc. Cambridge Philos. Soc. 62(1966), 379-387.

[26] N. Th. Varopoulos, Tensor algebra and harmonic analysis, Acta Math. 119(1967), 51-111.

[27] H. Whitney, On the extension of differentiable functions, Bull. Amer. Math. Soc. 50(1944), 76-81.

# SPECTRAL SYNTHESIS AND STABILITY IN SOBOLEV SPACES

Lars Inge Hedberg
University of Stockholm

## 0. Introduction and preliminaries

The approximation problems in Sobolev spaces which here will be called the spectral synthesis and stability problems arise in several different ways. The stability problem goes back to the fundamental work by M. V. Keldyš [23] on the stability of the solution of the Dirichlet problem and uniform approximation by harmonic functions. In greater generality the problem was raised by I. Babuška [4], who also noticed its connection with approximation in $L^2$ by harmonic and polyharmonic functions. V. P. Havin [17] proved that the problem of approximating in planar $L^2$ by rational functions is also equivalent to stability.

The closely related spectral synthesis problem in Sobolev spaces appears implicitly in the work of S. L. Sobolev [31], in the form of a uniqueness theorem for the Dirichlet problem for the polyharmonic equation $\Delta^m u = 0$. It is explicit in the work of Beurling and Deny [7], [11] on Dirichlet spaces. In more general Sobolev spaces it was formulated by Fuglede (see [30; IX, §5.1]). It also arises naturally in the context of $L^p$-approximation by analytic or harmonic functions, which was the motivation behind the work of T. Bagby [5], J. Polking [28], and the author [20], [22] on this problem.

The purpose of this paper is to explain the connection between these problems and to give a survey of work done on them in recent years, without going into too much technical detail. There are no new results in the paper; it is purely expository. It is in the nature of the problems treated here that the methods used are rather heavily potential-theoretical, but no previous knowledge of potential theory is assumed.

It has not been my purpose to write the history of the subject. This would be quite complicated, since results have often been discovered independently by different people. Neither has any attempt been made to give a complete bibliography. Several of the papers quoted contain comprehensive bibliographies, e.g. [18], [19], [22], and [26].

I am grateful to the University of Maryland and the University of California at Los Angeles for giving me the opportunity to give the series of lectures, of which these notes are a write-up.

Our notational conventions will be the following. The boundary of a set E is $\partial E$, its interior $E^0$, its closure $\overline{E}$, and its complement $E^c$. K is always going to be a compact set, usually in d-dimensional space $\mathbb{R}^d$, F is closed, and G is open. $C_0^\infty(G)$ means the $C^\infty$ functions with compact support in G. Various constants, whose value may change within the same sequence of inequalities, will be denoted by A.

The Sobolev space $W_m^p(\mathbb{R}^d) = W_m^p$ is defined as the closure of $C_0^\infty(\mathbb{R}^d)$ with respect to the norm $\|\cdot\|_{m,p}$ defined by

$$\|f\|_{m,p}^p = \sum_{0\le|\alpha|\le m} \int_{\mathbb{R}^d} |\partial^\alpha f(x)|^p \, dx.$$

Here $\alpha = (\alpha_1,\ldots,\alpha_d)$ is a multiindex of order $|\alpha| = \alpha_1+\ldots+\alpha_d$, $x = (x_1,\ldots,x_d)$, $dx = dx_1\ldots dx_d$ is Lebesgue measure, m is a positive integer, and $1 \le p < \infty$.

For any open $G \subset \mathbb{R}^d$ we denote by $\mathring{W}_m^p(G)$ the subspace of $W_m^p(\mathbb{R}^d)$ obtained as the closure of $C_0^\infty(G)$ with respect to $\|\cdot\|_{m,p}$.

Equivalently, $W_m^p$ can be defined as the functions in $L^p$, all of whose partial derivatives in the weak (or distribution) sense of order $\le m$ are also functions in $L^p$.

It is a basic fact that $\|f\|_\infty \le A\|f\|_{m,p}$, so $W_m^p(\mathbb{R}^d)$ can be continuously embedded in $C(\mathbb{R}^d)$, if $mp > d$, but not if $mp \le d$. The book by E. M. Stein [32] is a good reference for these and related facts.

In the following section we shall define and discuss spectral synthesis and stability in the classical context in $W_1^2$. In Section 2 we generalize to $W_1^p$, $p \neq 2$, and discuss how this leads to a non-linear potential theory. Finally, in Section 3 we study $W_m^p$, $m \geq 2$.

1. Spectral synthesis and stability in $W_1^2$.

Our starting point is the solution of the Dirichlet problem by the Dirichlet principle. Let $G \subset \mathbb{R}^d$ be a bounded open set, and let $f \in W_1^2(\mathbb{R}^d)$. Consider the extremal problem of finding $\inf\{\int|\nabla g|^2\, dx;\ f - g \in \mathring{W}_1^2(G)\}$. The set of competing functions g is convex and closed. By a standard argument using the parallelogram identity one shows that there exists a unique extremal, $f_G$, say. $f_G$ satisfies $\int \nabla f_G \cdot \nabla\varphi\, dx = 0$ for all $\varphi \in C_0^\infty(G)$, which by Weyl's lemma implies that $f_G \in C^\infty(G)$ and $\Delta f_G = 0$ in G. I.e., $f_G$ is the (generalized) solution of the Dirichlet problem with boundary data f in the sense that $f - f_G \in \mathring{W}_1^2(G)$. It is easily proved that this solution is unique.

Thus, if we use $\int\nabla f \cdot \nabla g\, dx$ as an inner product, we see that $W_1^2(\mathbb{R}^d)$ splits into two perpendicular subspaces, $W_1^2(\mathbb{R}^d) = \mathring{W}_1^2(G) \oplus D_1^2(G)$. Here $D_1^2(G) = \{f \in W_1^2(\mathbb{R}^d);\ f \text{ is harmonic in } G\}$.

We can now formulate the first version of the spectral synthesis and stability problems.

Consider potentials $U^\mu(x) = \int|x-y|^{2-d}\, d\mu(y)$, where $\mu$ is a signed measure. (For simplicity we assume that $d \geq 3$. For $d = 2$ the kernel $|x|^{2-d}$ is replaced by $\log \frac{1}{|x|}$.) The energy $I(\mu)$ is defined by $I(\mu) = \int U^\mu\, d\mu$, whenever $\int U^{|\mu|}\, d|\mu| < \infty$. By a classical formula $\int U^\mu\, d\mu = A \int |\nabla U^\mu|^2\, dx$. It is easily seen that potentials with finite energy are dense in $W_1^2(\mathbb{R}^d)$. In fact, all $C_0^\infty$ functions can be represented in this way.

If the support of $\mu$ does not intersect G, then $U^\mu$ is harmonic in G. The problem of spectral synthesis is: Are the potentials $U^\mu$

<u>which belong to</u> $D_1^2(G)$ <u>dense in</u> $D_1^2(G)$?

If this is the case, we shall say that <u>$G^c$ admits</u> (1,2)-<u>spectral synthesis</u>. A fundamental result, due to A. Beurling and J. Deny ([7], [11]) is that in fact all closed sets admit (1,2)-spectral synthesis. This will be proved below (Theorem 1.13.).

The problem of stability was raised and solved (in a somewhat different form) by M. V. Keldyš [23]. Let K be compact. Then the problem can be formulated: <u>Are the potentials</u> $U^\mu$ <u>with</u> supp $\mu \subset K^c$ <u>dense in</u> $D_1^2(K^0)$? I.e. can the measures in the spectral synthesis problem be pushed off the boundary $\partial K$? If this is the case, we say that <u>K is</u> (1.2)-<u>stable</u>. All compact sets are not (1,2)-stable, but Keldyš found a necessary and sufficient condition which will also be given below (Theorem 1.21).

We first give some equivalent formulations of the stability property. Let again $f \in W_1^2$ and a compact K be given. Then for every open G containing K we have $\int|\nabla f_{K0}|^2\,dx \geq \int|\nabla f_G|^2 dx$. In fact, $f_{K0} - f_G \in \mathring{W}_1^2(G)$. Set $\sup\{\int|\nabla f_G|^2\,dx;\ K \subset G,\ G \text{ open}\} = M$. Then we can find a sequence $\{G_n\}_1^\infty$, $K \subset G_n \subset \bar{G}_n \subset G_{n-1}$, so that the increasing sequence $\int|\nabla f_{G_n}|^2\,dx$ converges to M as $n \to \infty$. For $n > m$ we have $f - f_{G_n} \in \mathring{W}_1^2(G_m)$ and thus also $f - \frac{1}{2}(f_{G_n} + f_{G_m}) \in \mathring{W}_1^2(G_m)$. It follows that $\int|\nabla f_{G_n} + \nabla f_{G_m}|^2\,dx \geq 4\int|\nabla f_{G_m}|^2\,dx$, and by the parallelogram identity $\int|\nabla f_{G_n} - \nabla f_{G_m}|^2 dx = 2\int|\nabla f_{G_m}|^2 dx + 2\int|\nabla f_{G_n}|^2 dx - \int|\nabla f_{G_n} + \nabla f_{G_m}|^2 dx \leq 2\int|\nabla f_{G_n}\int|^2 dx - 2\int|\nabla f_{G_m}|^2 dx \to 0$, as $n,m \to \infty$. Thus the sequence $\{f_{G_n}\}_1^\infty$ is Cauchy in $W_1^2$, the limit $\lim_{n\to\infty} f_{G_n} = f_K$ exists, and $\int|\nabla f_K|^2 dx = M$. $f_K$ is harmonic in $K^0$ and called the exterior solution of the Dirichlet problem for K with boundary values f. The following proposition now follows easily.

<u>Proposition</u> 1.1: K is (1,2)-stable if and only if $f_K = f_{K0}$ for all $f \in W_1^2$.

K is thus (1,2)-stable if and only if all harmonic functions in $K^0$ that have extensions to $W_1^2(\mathbb{R}^d)$ can be approximated in $W_1^2$ by functions harmonic on neighborhoods of K.

We define $D_1^2(K) = \overline{\cup_G D_1^2(G)}$, where the union is taken over all open G containing K, and the closure is taken in $W_1^2(\mathbb{R}^d)$. Then it is easy to prove the following.

Proposition 1.2: K is (1,2)-stable if and only if $D_1^2(K) = D_1^2(K^0)$.

A function $g \in W_1^2$ is orthogonal to $D_1^2(K)$ if and only if $g \in \cap_G \mathring{W}_1^2(G)$, where the intersection is taken over all open G containing K. We denote the intersection by $\mathring{W}_1^2(K)$. It is easily seen that $\mathring{W}_1^2(K) = \{g \in W_1^2;\ g = 0 \text{ off } K\}$. We can thus write $W_1^2(\mathbb{R}^d) = \mathring{W}_1^2(K) \oplus D_1^2(K)$. The following proposition is immediate.

Proposition 1.3: K is (1,2)-stable if and only if $\mathring{W}_1^2(K) = \mathring{W}_1^2(K^0)$.

In other words, K is (1,2)-stable if and only if for every $g \in W_1^2$ such that $g = 0$ off K there is a sequence $\{\varphi_n\}_1^\infty$, $\varphi_n \in C_0^\infty(K^0)$, such that $\lim_{n\to\infty} \int |\nabla g - \nabla\varphi_n|^2\, dx = 0$.

Keldyš originally studied stability in terms of uniform convergence. It is a non-trivial fact, which it would take us too far to prove, that the definition of stability given here is equivalent to the one given by Keldyš. See Keldyš [23], Deny [10] and Landkof [24; Ch. V, §5]. If $f \in C(\mathbb{R}^d)$ we again let $f_G$ denote the solution of the Dirichlet problem in G with boundary values f. The result is the following. (Note that by Tietze's extension theorem any function in C(K) can be extended to $C(\mathbb{R}^d)$.)

Theorem 1.4: K is (1,2)-stable if and only if every $f \in C(\mathbb{R}^d)$ which is harmonic in $K^0$ can be uniformly approximated on K by functions $f_G$, where G is open and G contains K.

In the complex plane stability (and spectral synthesis) allow an interpretation in terms of analytic functions. If $K \subset C$ we denote by $L^p_a(K)$ the subspace of $L^p(K)$ consisting of functions analytic in $K^0$. By $R^p(K)$ we denote the closure in $L^p(K)$ of the rational functions with poles off K. By Runge's theorem $R^p(K)$ is also the closure in $L^p(K)$ of the functions which are analytic on some neighborhood of K. Clearly $R^p(K) \subset L^p_a(K)$. We write $x = x_1 + ix_2$, $\bar{x} = x_i - ix_2$, $dx = dx_1\, dx_2$.

Theorem 1.5. (V. P. Havin [17]): $R^2(K) = L^2_a(K)$ if and only if K is (1,2)-stable.

Proof: Let $g \in L^2(K)$. We can assume that $g(x) = 0$ for $x \notin K$. Then $\int_K fg\, dx = 0$ for all f in $L^2_a(K)$ if and only if $g = \frac{\partial\varphi}{\partial\bar{x}}$ for some $\varphi \in \overset{\circ}{W}{}^2_1(K^0)$. This is easy to prove using Weyl's lemma, i.e. the fact that $\int f \frac{\partial\varphi}{\partial\bar{x}}\, dx = 0$ for all $\varphi \in C_0^\infty(K^0)$ if and only if f is analytic in $K^0$.

For the same reason, if $\int_K fg\, dx = 0$ for all f in $L^2_a(G)$ for some $G \supset K$, it follows that $g = \frac{\partial\varphi}{\partial\bar{x}}$, $\varphi \in \overset{\circ}{W}{}^2_1(G)$. Thus $\int_K fg\, dx = 0$ for all f in $R^2(K)$ if and only if $g = \frac{\partial\varphi}{\partial\bar{x}}$, $\varphi \in \bigcap_{G \supset K} \overset{\circ}{W}{}^2_1(G) = W^2_1(K)$. Thus $R^2(K) = L^2_a(K)$ if K is (1,2)-stable. Conversely, if $R^2(K) = L^2_a(K)$, then K is (1,2)-stable. In fact, let $\varphi \in \overset{\circ}{W}{}^2_1(K)$, and let $g = \frac{\partial\varphi}{\partial\bar{x}}$. Then, if $R^2(K) = L^2_a(K)$, we also have $g = \frac{\partial\psi}{\partial\bar{x}}$, where $\psi \in \overset{\circ}{W}{}^2_1(K^0)$. But then $\frac{\partial}{\partial\bar{x}}(\varphi-\psi) = 0$, and $\varphi-\psi$ has compact support, so $\varphi = \psi$, and thus $\varphi \in \overset{\circ}{W}{}^2_1(K^0)$.

In order to get further we have to define capacities. The functions in $W^2_1(\mathbb{R}^d)$ are not in general continuous if $d \geq 2$. ($f(x) = \log\log|x|$ for $|x| < e$, 0 for $|x| \geq e$ is an example in $\mathbb{R}^2$.) The natural way of measuring by how much the functions deviate from continuity is by means of capacity.

More generally, to every Sobolev space $W^p_m(\mathbb{R}^d)$ we associate an (m,p)-capacity by the following definition.

Definition 1.6: (a) If K is compact, $C_{m,p}(K) = \inf\{\|\omega\|^p_{m,p};\ \omega \in C_0^\infty,\ \omega \geq 1 \text{ on } K\}$.

(b) If G is open, $C_{m,p}(G) = \sup\{C_{m,p}(K);\ K \subset G,\ K \text{ compact}\}$.

(c) E is arbitrary, $C_{m,p}(E) = \inf\{C_{m,p}(G);\ G \supset E,\ G \text{ open}\}$.

A property is said to hold (m,p)-quasi everywhere ((m,p)-q.e.) if it is true for all x except those belonging to a set with zero (m,p)-capacity.

Let $f \in W^p_m \cap C$. Then the following inequality is an immediate consequence of the definition of capacity.

$$C_{m,p}(\{x;|f(x)| > \lambda\} \leq \frac{1}{\lambda^p} \|f\|^p_{m,p}, \quad \lambda > 0.$$

One can prove a similar inequality for the Hardy-Littlewood maximal function Mf, defined by

$$Mf(x) = \sup_{r>0} \frac{1}{r^d} \int_{|y-x|\leq r} |f(y)|dy.$$

Theorem 1.7: (D. R. Adams [1]). Let $f \in W^p_m$. Then $C_{m,p}(\{x;Mf(x) > \lambda\}) \leq \frac{A}{\lambda^p} \|f\|^p_{m,p}$, $\lambda > 0$.

Now let $\chi \in C_0^\infty(\{|x| < 1\})$, $\chi \geq 0$, and $\int \chi\, dx = 1$. Define an approximate identity $\{\chi_n\}_1^\infty$, by $\chi_n(x) = n^d\chi(nx)$. If $f \in W^p_m$ if follows that $f * \chi_n \in C^\infty$ and that $\|f - f * \chi_n\|_{m,p} \to 0$ as $n \to \infty$. Using Theorem 1.7, and standard arguments one also proves that $\lim_{n\to\infty} f * \chi_n(x) = \tilde{f}(x)$ exists for (m,p) - q.e. x. Clearly $f(x) = \tilde{f}(x)$ a.e. Moreover, there is a subsequence $\{\chi_{n_k}\}_{k=1}^\infty$ such that for any given $\varepsilon > 0$ there is an open G, $C_{m,p}(G) < \varepsilon$, such that $f * \chi_{n_k} \to \tilde{f}(x)$ uniformly outside G. Thus $\tilde{f}|_{G^c}$ is continuous on $G^c$. A function which is defined (m,p) - q.e. and has this continuity property is called (m,p)-quasicontinuous.

Thus we can extend the definition of f by setting $f(x) = \tilde{f}(x)$. Moreover, this extension is essentially unique. We summarize the result. (See Deny-Lions [12], Wallin [33], Havin-Maz'ja [18]).

Theorem 1.8: Let $f \in W_m^p$. Then, after possible redefinition on a set of measure zero, f is (m,p)-quasicontinuous. Moreover, if f and g are two (m,p)-quasicontinuous functions such that $f(x) = g(x)$ almost everywhere, then $f(x) = g(x)$ (m,p)-q.e.

In what follows functions in $W_m^p$ are always assumed (m,p)-quasicontinuous. It then also makes sense to talk about the restriction or trace of functions in $W_m^p$ on arbitrary sets of positive capacity. Thus if we write $f|_F = 0$ for a function f in $W_m^p$, this means that $f(x) = 0$ (m,p)-q.e. on F.

If $mp < d$ one gets the same nullsets if (m,p)-capacity is defined using only derivatives of order m. For example, if $d \geq 3$, one can define $C_{1,2}(K) = \inf\{\int |\nabla\omega|^2\, dx;\ \omega \in C_0^\infty,\ \omega \geq 1 \text{ on } K\}$. (We omit the modifications necessary if $d = 2$.)

By classical theorems of Frostman, there is then a unique extremal $\omega_K$, and a positive measure $\mu_K$ with support in K and $\mu_K(K) = C_{1,2}(K)$, such that

$$\omega_K(x) = \int \frac{d\mu_K(y)}{|x-y|^{d-2}} = U^{\mu_K}(x).$$

The results extend to arbitrary sets.

Theorem 1.9: For any bounded $E \subset \mathbb{R}^d$ there exists a unique measure $\mu_E \geq 0$ with support in $\bar{E}$ such that

(a) $U^{\mu_E}(x) = 1$ (1,2)-q.e. on E;

(b) $U^{\mu_E}(x) \leq 1$ for all x; (Note that $U^{\mu_E}(x)$ is defined everywhere.)

(c) $\int_{\mathbb{R}^d} d\mu_E = \int U^{\mu_E}\, d\mu_E = I(\mu_E) = C_{1,2}(E)$.

$U^{\mu_E}$ is called the equilibrium or capacitary potential for E. See e.g. Landkof [24].

E is said to be thin (or (1,2)-thin) at the points where $U^{\mu_E}(x) < 1$. More precisely, we define thinness in the following way.

Definition 1.10: A set E is (1,2)-thin at x if either $x \notin \bar{E}$ or $x \in \bar{E}$ and there exists a positive measure $\mu$ such that

$$U^{\mu}(x) < \liminf_{y \to x,\, y \in E \setminus \{x\}} U^{\mu}(y).$$

A necessary and sufficient condition for thinness is given by the Wiener criterion. See [24].

Theorem 1.11: A set $E \subset \mathbb{R}^d$ is (1,2)-thin at x if and only if

(a) $\sum_{n=1}^{\infty} 2^{n(d-2)} C_{1,2}(E \cap A_n(x)) < \infty, \quad d \geq 3$

(b) $\sum_{n=1}^{\infty} n\, C_{1,2}(E \cap A_n(x)) < \infty, \quad d = 2.$

Here $A_n(x)$ is the annulus $\{y; 2^{-n-1} \leq |y-x| < 2^{-n}\}$.

If a set is not thin we shall say that it is thick. Any set is (1,2)-thick at all interior points. In particular $U^{\mu_E}(x) = 1$ everywhere in $E^0$.

We can now give a dual interpretation of spectral synthesis.

Theorem 1.12: Let $G \subset \mathbb{R}^d$ be open and bounded. Potentials $U^{\mu}$ belonging to $D_1^2(G)$ are dense in $D_1^2(G)$ if and only if every f in $W_1^2$ such that $f|_{G^c} = 0$ belongs to $\overset{\circ}{W}{}_1^2(G)$.

Proof: We know that the orthogonal complement to $D_1^2(G)$ is $\overset{\circ}{W}{}_1^2(G)$ In order to prove the theorem it is enough to show that the orthogonal complement of $\{U^{\mu};\, U^{\mu} \in D_1^2(G)\}$ consists of the functions $f \in W_1^2$ such that $f|_{G^c} = 0$.

Let $f \in W_1^2$. Formally

$$\int \nabla f \cdot \nabla U^{\mu}\, dx = \int \left( \nabla f(x) \cdot \int \frac{x-y}{|x-y|^d}\, d\mu(y) \right) dx$$

$$= \iint \frac{x-y}{|x-y|^d} \cdot \nabla f(x) dx\, d\mu(y)$$

$$= A \int f(y) d\mu(y).$$

But by a classical identity of M. Riesz

$$I(\mu) = \int U^\mu \, d\mu = A \int \left\{ \int \frac{d\mu(y)}{|x-y|^{d-1}} \right\}^2 dx.$$

It follows from Fubini's theorem that the above change of order of integration is justified if $I(\mu) < \infty$.

Using the regularizing sequence $\{f_n\} = \{f * \chi_n\}$ it is now easy to show that

$$\int f d\mu = \lim_{n\to\infty} \int f_n \, d\mu = \lim_{n\to\infty} A \int \nabla f_n \cdot \nabla U^\mu \, dx = A \int \nabla f \cdot \nabla U^\mu \, dx$$

for all $f \in W_1^2$ and all $\mu$ with $I(\mu) < \infty$. The theorem follows easily.

If F is an arbitrary closed set we take the property in Theorem 1.12 as our definition of (1,2)-spectral synthesis. (If we had chosen to consider the equation $-\Delta u + u = 0$ instead of $\Delta u = 0$ the restriction to bounded G would not have been necessary.)

We can now prove the theorem of Beurling and Deny referred to earlier.

Theorem 1.13: Every closed set in $\mathbb{R}^d$ admits (1,2)-spectral synthesis.

Proof: The crucial property of $W_1^2$ is that this space is closed under truncations. For example, if $f \in W_1^2$, then $f^+ = \max(f,0) \in W_1^2$, and $\int |\nabla f^+|^2 \, dx = \int_{\{f(x)>0\}} |\nabla f|^2 \, dx \le \int |\nabla f|^2 \, dx$.

Let $f \in W_1^2$, and suppose that $f|_F = 0$ for a closed set F. We claim that f can be approximated by functions that vanish on a neighborhood of F. By truncating and multiplying by a cut-off function we can always assume that f is bounded and has compact support. It is sufficient to consider $f^+$.

Let $\varepsilon > 0$ and let $f_\varepsilon = (f^+ - \varepsilon)^+$. If f is continuous, then $f_\varepsilon(x) = 0$ in a neighborhood of F, and

$$\int |\nabla f - \nabla f_\varepsilon|^2 \, dx = \int_{0<f^+<\varepsilon} |\nabla f|^2 \, dx \to 0, \ \varepsilon \to 0,$$

which proves the theorem in this case.

In general f is only quasicontinuous. Then, for any n there is an open set $G_n$, $C_{1,2}(G_n) < \frac{1}{n}$, such that $f|_{G_n^c}$ is continuous on $G_n^c$, and such that $f(x) = 0$ on $F\setminus G_n$. There is an $\omega_n \in W_1^2$ such that $\omega_n(x) = 1$ on $G_n$, $\int|\nabla\omega_n|^2\,dx < \frac{1}{n}$, and $0 < \omega_n(x) \leq 1$. Using the boundedness of f one easily shows that $f_\varepsilon(1-\omega_n)$ is the required approximation to $f^+$. (See [20].)

In the complex plane the theorem has a dual interpretation analogous to Theorem 1.5. By the Cauchy transform of a measure we mean the function $\hat{\mu}(x) = \int(\xi-x)^{-1}\,d\mu(\xi)$, $x = x_1 + ix_2$.

<u>Corollary</u> 1.14: ([20]). For any bounded open $G \subset C$ the Cauchy transform $\hat{\mu}$ of measures with support in $G^c$ are dense in $L_a^2(G)$.

Returning to stability we observe the following immediate corollary to Theorem 1.13.

<u>Corollary</u> 1.15: K is (1,2)-stable if and only if every $f \in \mathring{W}_1^2(K)$ satisfies $f|_{\partial K} = 0$.

In fact, by the theorem $\mathring{W}_1^2(K^0)$ consists exactly of the functions f with $f|_{(K^0)^c} = 0$.

We can now give a number of necessary and sufficient conditions for a set K to be stable.

<u>Theorem</u> 1.16: (T. Bagby [5])

(a) K is (1,2)-stable if and only if $C_{1,2}(G\setminus K) = C_{1,2}(G\setminus K^0)$ for all open G.

(b) K is (1,2)-stable if and only if for some $\eta > 0$ $C_{1,2}(G\setminus K) \geq \eta\, C_{1,2}(G\setminus K^0)$ for all open G.

One proves in one direction that the property in (b) extends to all so called quasi-open sets G, in particular to the set $G = \{x; |f(x)|>0\}$ for any $f \in W_1^2$. Now suppose $f \in \mathring{W}_1^2(K)$, i.e.

$f(x) = 0$ on $K^c$. It follows that $C_{1,2}(G\setminus K^0) \le \eta^{-1} C_{1,2}(G\setminus K) = 0$. i.e., $|f(x)| = 0(1,2)$-q.e. on $\partial K$, q.e.d.

The other direction, that stability implies (a) follows easily from the definitions. See [5], [20].

Using this theorem one easily gets an example of an unstable set.

Example 1.17: Let $B_0$ be the open unit disk in the plane, and let $I = [-\frac{1}{2},\frac{1}{2}]$ on the real axis. Let $\{B_i\}^\infty$, be disjoint disks in $B_0\setminus I$ such that the disks accumulate at all points of $I$ but nowhere else. Let $K = \bar{B}_0\setminus(\cup_1^\infty B_i)$. Then $\partial K = I\cup(\cup_0^\infty B_i)$. If the $B_i$ are small enough $K$ will be unstable. In fact, $C_{1,2}(B_0\setminus K^0) \ge C_{1,2}(I) > 0$, as is well known. On the other hand

$$C_{1,2}(B_0\setminus K) = C_{1,2}(\bigcup_1^\infty B_i) \le \sum_1^\infty C_{1,2}(B_i),$$

which can be made arbitrarily small.

Theorem 1.18: (Havin [17]) $K$ is (1,2)-stable if and only if $K^c$ is (1,2)-thick (1,2)-q.e. on $\partial K$.

Proof: If $K^c$ is thick q.e. on $\partial K$, then it follows easily from the definitions of thinness and of capacity that $C_{1,2}(G\setminus K) = C_{1,2}(G\setminus K^0)$ for all open $G$. In fact, the capacitary potential for $G\setminus K$ is 1 (1,2)q.e. on $G\setminus K^0$.

In the converse direction the proof depends on a lemma of Choquet [9]. For any set $E$ we let $e(E)$ denote the set of points where $E$ is thin. Thus $e(E)$ contains the exterior and part of the boundary of $E$.

Lemma 1.19: Let $E$ be an arbitrary set. For any $\varepsilon > 0$ there exists an open $G$ such that $e(E) \subset G$ and $C_{1,2}(G \cap E) < \varepsilon$.

The Kellogg property is an immediate consequence.

Corollary 1.20: For any $E$, $C_{1,2}(E \cap e(E)) = 0$.

To prove Theorem 1.18 we let A be the subset of $\partial K$ where $K^c$ is thin, and suppose $C_{1,2}(A) > \varepsilon > 0$. By Lemma 1.19 applied to $E = K^c$ there is an open G such that $G \supset E$, and $C_{1,2}(G\backslash K) < \varepsilon$. But $C_{1,2}(G\backslash K^0) \geq C_{1,2}(A) > \varepsilon$, so $C_{1,2}(G\backslash K) < C_{1,2}(G\backslash K^0)$, and thus K is not (1,2)-stable by Theorem 1.16. (In terms of the fine topology of potential theory Theorem 1.18 says that K is (1,2)-stable if and only if the part of the fine interior of K which belongs to $\partial K$ has capacity zero.)

<u>Theorem</u> 1.21: (Keldyš [23]). K is (1,2)-stable if and only if $K^c$ and $(K^0)^c$ are (1,2)-thin at the same points.

<u>Proof</u>: If $C_{1,2}(G\backslash K) = C_{1,2}(G\backslash K^0)$ for all open G, then the Wiener series for $K^c$ and $(K^0)^c$ converge simultaneously. In the other direction the condition of the theorem implies that $C_{1,2}(G\backslash K) = C_{1,2}(G\backslash K^0)$ as in Theorem 1.18.

<u>Theorem</u> 1.22: K is (1,2)-stable if and only if

$$\liminf_{r\to 0} \frac{C_{1,2}(B(x,r)\backslash K)}{C_{1,2}(B(x,r)\backslash K^0)} > 0 \quad \text{for (1,2)-q.e.} \quad x \in \partial K.$$

Here $B(x,r) = \{y; |y-x| < r\}$.

<u>Proof</u>: Assume that K satisfies the above lim inf condition. By Kellogg's lemma (Cor. 1.20) the Wiener series for $(K^0)^c$ diverges (1,2)-q.e. on $\partial K$. But then by the assumption the Wiener series for $K^c$ diverges, so $K^c$ is thick q.e. on $\partial K$. The theorem follows from Theorem 1.18. The other direction is obvious.

If $K^0 = \emptyset$ these results can be sharpened.

<u>Theorem</u> 1.23: (Gončar [16], Lysenko and Pisarevskiĭ [25]). Let $K^0 = \emptyset$. Then K is (1,2)-stable if either

(a) $C_{1,2}(G\backslash K) = C_{1,2}(G)$ for all open G

or

(b) $\limsup_{r\to 0} C_{1,2}(B(x,r)\backslash K)r^{-d} > 0$ for a.e. x.

The phenomenon that (a) and (b) are equivalent, although $C_{1,2}(B(x,r) \approx r^{d-2}$, is called the "instability of capacity". See also [20] and Fernström [13].

2. Generalization to $W_1^p$, $p \neq 2$.

It is natural to try to generalize the above to $p \neq 2$, and investigate conditions for e.g. $R^p(K) = L_a^p(K)$. One is led to the following definitions. We let $1 < p < \infty$, $\frac{1}{p} + \frac{1}{q} = 1$.

Definition 2.1: A closed $F \subset \mathbb{R}^d$ admits $(1,p)$-spectral synthesis if all $f \in W_1^p(\mathbb{R}^d)$ such that $f|_F = 0$ belong to $\mathring{W}_1^p(F^c)$.

Definition 2.2: A compact $K \subset \mathbb{R}^d$ is $(1,p)$-stable if $\mathring{W}_1^p(K^0) = \mathring{W}_1^p(K)$ $(= \cap_{G \supset K} \mathring{W}_1^p(G))$.

$W_1^p$ is closed under truncation for all p, so the proof of the Beurling-Deny theorem given above extends almost unchanged.

Theorem 2.3: (T. Bagby [5]) All closed sets admit $(1,p)$-spectral synthesis, $1 < p < \infty$.

Corollary 2.4: ([20]) For any open bounded $G \subset \mathbb{C}$ the Cauchy transforms $\hat{\mu}$ of measures $\mu$ such that $\hat{\mu} \in L_a^q(G)$ are dense in $L_a^q(G)$, $1 < q < \infty$.

Remark: This corollary is true for $q = 1$ also, but this requires a different proof, due to L. Bers [6].

The dual space to $W_1^p$ is denoted $W_{-1}^q$. It consists of those distributions which have continuous extensions to $W_1^p$. The elements can be represented as linear combinations of first derivatives of $L^q$-functions. In terms of $W_{-1}^q$ Definition 2.1 can easily be given an equivalent dual formulation, which leads to the following corollary.

Corollary 2.5: For any closed $F \subset \mathbb{R}^d$ all $T \in W_{-1}^q$, $1 < q < \infty$, with supp $T \subset F$ can be approximated in $W_{-1}^q$ by measures $\mu$ with supp $\mu \subset F$.

The following two theorems are proved much as Theorems 1.5 and 1.6.

Theorem 2.6 (Havin [17]): Let $K \subset \mathbb{C}$. Then $R^q(K) = L_a^q(K)$, $1 < q < \infty$, if and only if K is (1,p)-stable.

Theorem 2.7 (Havin [17], Bagby [5]): All $K \subset \mathbb{R}^d$ are (1,p)-stable if $p > d$. If $1 < p < d$, K is (1,p)-stable if and only if either

(a) $C_{1,p}(G \setminus K) = C_{1,p}(G \setminus K^0)$ for all open G

(b) there is an $\eta > 0$ such that $C_{1,p}(G \setminus K) \geq \eta\, C_{1,p}(G \setminus K^0)$ for all open G.

If one tries to generalize Theorems 1.18, 1.21 and 1.22 one immediately runs into problems. What is a proper generalization of (1,2)-thinness? In the case p = 2 the extremal $\omega_K$ is harmonic on $K^c$, and is represented as a potential of a positive measure. For $p \neq 2$ $\omega_K$ satisfies the much more complicated equation $\operatorname{div}(|\nabla\omega_K|^{p-2}\, \nabla\omega_K) = 0$, and it cannot be represented as a potential.

Fortunately, there is a different approach to the potential theory of $W_m^p$, which originated in works of B. Fuglede [15] and N. G. Meyers [27].

If $f \in W_m^p$, $m < d$, then f can be represented as a Riesz potential

$$f(x) = \int \frac{g(y)\,dy}{|x-y|^{d-m}} = R_m * g,$$

where $g \in L^p$. However, if $g \in L^p$, then $R_m * g$ is not necessarily in $W_m^p$. The reason for this is that $R_m(x)$ tends to zero too slowly at $\infty$. This corresponds to the fact that the Fourier transform $\hat{R}_m(\xi) = A|\xi|^{-m}$ has a singularity at 0. Therefore, for any $\alpha > 0$ one defines the so called Bessel kernel $G_\alpha(x)$ as the inverse Fourier transform of $\hat{G}_\alpha(\xi) = (1 + |\xi|^2)^{-\alpha/2}$. In other words, $G_\alpha * g = (I-\Delta)^{-\alpha} g$. The Bessel kernels have the following properties. See e.g. Stein [32].

(a) $G_\alpha(x) \geq 0$;

(b) $G_\alpha * G_\beta = G_{\alpha+\beta}$;

(c) $G_\alpha(x) = O(e^{-c|x|})$, $|x| \to \infty$;

(d) $G_\alpha(x) \approx A|x|^{\alpha-d}$, $|x| \to 0$, $0 < \alpha < d$;

(e) $G_d(x) \approx A \log \frac{1}{|x|}$, $|x| \to 0$.

Using the theory of singular integrals the following theorem is now easy to prove.

Theorem 2.8 (Calderón [8]): $f \in W_m^p$, $1 < p < \infty$, if and only if $f = G_m * g$, where $g \in L^p$. Moreover, there is a constant $A > 0$ such that $A^{-1}\|g\|_p \leq \|f\|_{m,p} \leq A\|g\|_p$.

Definition 2.9: The Bessel potential space $L_\alpha^p(\mathbb{R}^d) = \{G_\alpha * g; g \in L^p(\mathbb{R}^d)\}$, and $\|f\|_{\alpha,p} = \|g\|_p$.

We now modify the definition of capacity by using $L_\alpha^p$.

Definition 2.10: For any set $E \subset \mathbb{R}^d$ the $(\alpha,p)$-capacity

$$C_{\alpha,p}(E) = \inf\left\{\int g^p\,dx; g \geq 0,\ G_\alpha * g(x) \geq 1 \quad \text{on } E\right\}.$$

Note that the definition makes sense for arbitrary E, since $G_\alpha * g(x)$ is defined everywhere for $g \geq 0$. Again, $G_\alpha * g$ is continuous for $\alpha p > d$, $C_{\alpha,p}(E) > 0$ for all non-empty sets, and the capacity is not interesting for our purposes.

We shall investigate the extremal function in the definition of capacity. Let K be compact, and let $\mu$ be a positive measure with support on K. Let $g \geq 0$ and $G_\alpha * g \geq 1$ on K. Then by Fubini's theorem

$$\mu(K) \leq \int (G_\alpha * g)\,d\mu = \int (G_\alpha * \mu)g\,dx \leq \|G_\alpha * \mu\|_q\,\|g\|_p.$$

Thus
$$\sup_\mu \frac{\mu(K)}{\|G_\alpha * g\|_q} \leq \inf_g \|g\|_p = C_{\alpha,p}(K)^{1/p}.$$

Applying the Minimax Theorem to the bilinear functional $\Phi(g,\mu) = \int (G_\alpha * \mu)g\,dx$ one can show that equality holds in the last inequality.

(Fuglede [15], Meyers [27]). Moreover there are extremal $\mu_K$ and $g_K$ so that $G_\alpha * g_K \geq 1$ $(\alpha,p)$-q.e. on K, and

$$\mu_K(K) = \int (G_\alpha * \mu_K)g_K\, dx = \|G_\alpha * \mu_K\|_q\, \|g_K\|_p.$$

It follows that $(G_\alpha * \mu_K)^q = Ag_K^p$. Choosing A = 1 we have $g_K = (G_\alpha * \mu_K)^{q-1}$, and

$$\mu_K(K) = \int (G_\alpha * \mu_K)^q\, dx = \int g_k^p\, dx = C_{\alpha,p}(K).$$

The function $G_\alpha * g_K = G_\alpha * (G_\alpha * \mu_K)^{q-1} = V_{\alpha,p}^{\mu_K}$ is called a <u>non-linear potential</u> of the measure $\mu_K$. (If p = 2 the non-linearity disappears; $V_{\alpha,2}^{\mu} = G_\alpha * G_\alpha * \mu = G_{2\alpha} * \mu$, which is a classical potential.)

The results extend to arbitrary sets (See [27]). We summarize.

<u>Theorem</u> 2.11: For every bounded $E \subset \mathbb{R}^d$, $1 < p < \infty$, $\alpha > 0$, there is a unique measure $\mu_E \geq 0$, the capacitary measure, with support in $\bar{E}$ such that

(a) $V_{\alpha,p}^{\mu_E}(x) \geq 1$ $(\alpha,p)$-q.e. on E;

(b) $V_{\alpha,p}^{\mu_E}(x) \leq 1$ for all $x \in \operatorname{supp} \mu_E$;

(c) $\int_{\mathbb{R}^d} d\mu_E = \int V_{\alpha,p}^{\mu_E}\, d\mu_E = \int (G_\alpha * \mu_E)^q\, dx = C_{\alpha,p}(E)$.

It is easy to see that $C_{\alpha,p}(E) = \inf\{C_{\alpha,p}(G); G \supset E,\ G \text{ open}\}$ for any E. One can show that $C_{\alpha,p}$ satisfies the axioms of Choquet's theory of capacitability (Fuglede [15], Meyers [27]). Thus

<u>Theorem</u> 2.12: For every Borel (or Suslin) set E $C_{\alpha,p}(E) = \sup\{C_{\alpha,p}(K); K \subset E,\ K \text{ compact}\}$.

In general $V_{\alpha,p}^{\mu_E}(x) > 1$ on $E^0$. This is for example the case if p = 2 and $\alpha > 2$. However, one can prove the following "boundedness principle". (Havin-Maz'ja [18], Adams-Meyers [3]).

<u>Theorem</u> 2.13: Let $\mu > 0$. There is a constant A, only depending on d and p, such that for all x

$$V_{\alpha,p}^{\mu}(x) \leq A \max\{V_{\alpha,p}^{\mu}(y); y \in \operatorname{supp} \mu\}.$$

Thus in particular, the capacitary potential $V_{\alpha,p}^{\mu_E}$ is bounded by A.

The theory of non-linear potentials was studied systematically by V. P. Havin and V. G. Maz'ja and they gave many applications to various problems in analysis. (See [18], [19].) At the same time several of their results were found independently by N. G. Meyers and D. R. Adams [2], [3].

The following natural extension of the definition of thinness was given by Adams and Meyers [2] and, independently, by the author [20].

Definition 2.14: A set E is $(\alpha,p)$-thin at x if either $x \notin \bar{E}$ or $x \in \bar{E}$ and there is a positive measure $\mu$ such that

(a) $V_{\alpha,p}^{\mu}$ is bounded;

(b) $V_{\alpha,p}^{\mu}(x) < \liminf_{y\to x,\, y\in E\setminus\{x\}} V_{\alpha,p}^{\mu}(y)$.

Many of the properties of (1,2)-thin sets extend to this more general setting. See [2]. The following is a special case of a theorem of Fuglede [14]. See also [20].

Theorem 2.15: Let $f \in L_{\alpha}^{p}$. For $(\alpha,p)$-q.e. x the set $E_\lambda = \{y; f(y) \text{ is not defined or } |f(y)-f(x)| \geq \lambda\}$ is $(\alpha,p)$-thin at x for all $\lambda > 0$.

A problem which has not yet found a satisfactory solution is the generalization of Wiener's criterion. The following is known ([2], and in part [20]).

Set $2^{n(d-\alpha p)} C_{\alpha,p}(E \cap B(x,2^{-n})) = a_n(x,E)$. Note that if E is a cone with vertex at x, then $\lim_{n\to\infty} a_n(x,E)$ is finite and positive for $0 < \alpha p < d$.

Theorem 2.16: (a) If $\sum_1^\infty a_n(x,E)^{q-1} = \infty$, then E is $(\alpha,p)$-thick at x, $1 < p < \infty$, $o < \alpha p \leq d$.

(b) If $\sum_{n=1}^{\infty} a_n(x,E)^{q-1} < \infty$, $p > 2 - \frac{\alpha}{d}$, then E is $(\alpha,p)$-thin at x.

(c) If $\sum_{n=1}^{\infty} a_n(x,E)^{(d-\alpha)/(d-\alpha p)} < \infty$, $1 < p < 2 - \frac{\alpha}{d}$, then E is $(\alpha,p)$-thin at x.

(d) If $\sum_{n=1}^{\infty} \left\{ a_n(x,E) \log \frac{A}{a_n(x,E)} \right\}^{q-1} < \infty$, $p = 2 - \frac{\alpha}{d}$, then E is $(\alpha,p)$-thin at x.

It is also known ([2]) that the exponent in (c) is best possible. The number $2 - \frac{\alpha}{d}$ appears because the estimate which leads to (b) breaks down when $G_\alpha$ no longer belongs to $L^{q-1}$. For $p = 2$ the above theorem contains the Wiener criterion. (See [24]).

Using the estimates involved in proving Theorem 2.16 one can now easily extend Choquet's lemma in the case $p > 2 - \frac{\alpha}{d}$, and as a corollary the Kellogg property. Whether the Choquet and Kellogg properties are true for $1 < p \leq 2 - \frac{\alpha}{d}$ remains unknown.

Putting these results and the spectral synthesis theorem together one obtains the following generalization of Theorem 1.18.

Theorem 2.17: A compact $K \subset \mathbb{R}^d$ is $(1,p)$-stable if $K^c$ is $(1,p)$-thick $(1,p)$-q.e. on $\partial K$.

If $p > 2 - \frac{1}{d}$, then the converse is also true.

Theorems 1.21 and 1.22 generalize in the same way. Theorem 1.23 generalizes without restriction. (See [20]).

## 3. Higher derivatives.

What should be meant by $(m,p)$-spectral synthesis if $m > 1$? Let $F \subset \mathbb{R}^d$ be a closed set. If a function f in $W_m^p(\mathbb{R}^d)$, $m > 1$, is to belong to $\mathring{W}_m^p(F^c)$ it is clear that one has to require more than that $f|_F = 0$. In fact, if $\varphi_n \in C_0^\infty(F^c)$, $\varphi_n \to f$ in $W_m^p$, then $\nabla\varphi_n \to \nabla f$ in $W_{m-1}^p$, $\nabla^2\varphi_n \to \nabla^2 f$ in $W_{m-2}^p$, etc. Thus $\nabla^k f(x) = 0$ $(m-k,p)$-q.e. on F, i.e. $\nabla^k f|_F = 0$ for $0 \leq k \leq m-1$. (Here $\nabla^k f$ means the vector valued function consisting of all partial derivatives of f of order k.)

Definition 3.1: (Fuglede, see [30; IX,§5] and [22]). F admits (m,p)-spectral synthesis if every $f \in W^p_m$ such that $\nabla^k f|_F = 0$, $0 < k \le m - 1$, belongs to $\mathring{W}^p_m(F^c)$.

In other words, F admits (m,p)-spectral synthesis if the obvious necessary condition for f to belong to $\mathring{W}^p_m(F^c)$ is also sufficient.

Conjecture: All closed $F \subset \mathbb{R}^d$ admit (m,p)-spectral synthesis for $1 < p < \infty$, $m = 1,2,\ldots$ .

The reason why this is a problem for $m > 1$ is that the spaces $W^p_m$ are no longer closed under truncation. Even if $f \in C^\infty_0$, the distribution derivatives of $\nabla(f^+)$ are not, in general, functions.

Some positive results will be given below.

The concept of stability can also be generalized in a natural way. As before we write

$$\mathring{W}^p_m(K) = \bigcap_{G \supset K} \mathring{W}^p_m(G) = \{f \in W^p_m;\ f(x) = 0 \quad \text{on } K^c\}.$$

Definition 3.2 (Babuška [4]): $K \subset \mathbb{R}^d$ is called (m,p)-stable if $\mathring{W}^p_m(K) = \mathring{W}^p_m(K^0)$.

Clearly, K can be (m,p)-stable only if $(K^0)^c$ admits (m,p)-spectral synthesis. General necessary and sufficient conditions for (m,p)-stability are not known, but some sufficient conditions will be given below.

Again, the spectral synthesis and stability properties can be given several equivalent formulations. For simplicity of statement we let $m = 2$.

Thus, F admits (2,p)-spectral synthesis if and only if every distribution T in the dual space $W^q_{-2}$ with supp $T \subset F$ can be approximated in $W^q_{-2}$ by linear combinations $\mu_0 + \sum_1^d \partial\mu_i/\partial x_i$, where $\mu_i$ are measures with supp $\mu_i \subset F$, $\mu_0 \in W^q_{-2}$, $\mu_1,\ldots,\mu_d \in W^q_{-1}$.

For any E we let $L^q_h(E)$ denote the subspace of $L^q(E)$ consisting of functions harmonic in $E^0$. If G is open and bounded, then $G^c$ admits (2,p)-spectral synthesis if and only if every $f \in L^q_h(G)$ can be approximated in $L^q(G)$ by linear combinations $U^{\mu_0} + \sum_1^d \partial U^{\mu_i}/\partial x_i$, where $\mu_i$ are measures with support in $G^c$, $\mu_0 \in W^q_{-2}$, $\mu_1,\ldots,\mu_d \in W^q_{-1}$.

It is easily seen directly from examples that one cannot in general expect potentials $U^\mu$, supp $\mu \subset G^c$, to be dense in $L^q_h(G)$. If for example $G = \{0 < |x| < 1\}$ in $\mathbb{R}^2$ and $q < 2$, then $f(x) = x_1/(x_1^2 + x_2^2) \in L^q_h(G)$, but the only potentials with a singularity at 0 are multiples of $\log 1/|x|$.

Similarly, a compact set is (2,p)-stable if and only if all $f \in L^q_h(K)$ can be approximated in $L^q(K)$ by functions harmonic on neighborhoods of K.

(2,2)-spectral synthesis and stability are closely related to the Dirichlet problem for the biharmonic equation, $\Delta^2 u = 0$.

If G is a bounded domain with smooth boundary, and if f and g are two continuous functions, given on $\partial G$, then the Dirichlet problem is to find a function $u \in C^1(\bar{G}) \cap C^4(G)$ such that $\Delta^2 u = 0$ in G, $u|_{\partial G} = f$ and the normal derivative $\partial u/\partial n = g$ on $\partial G$.

In the general case when G does not have a regular boundary the problem has to be formulated differently. One standard way is to copy the procedure we used for the Laplace equation. Let G be open and bounded, but otherwise arbitrary, and let f be a given function which we assume belongs to $W^2_2(\mathbb{R}^d)$. Then the Dirichlet problem is to find a function $u \in W^2_2(\mathbb{R}^d)$ such that $\Delta^2 u = 0$ in G and $u - f \in \overset{\circ}{W}{}^2_2(G)$.

The solution can again be found by solving an extremal problem, in this case to find $\inf\{\int|\Delta g|^2\,dx;\ g \in W^2_2(\mathbb{R}^d),\ g - f \in \overset{\circ}{W}{}^2_2(G)\}$. The unique extremal, which we again denote $f_G$, is found by projection of f to $D^2_2(G)$, the orthogonal complement of $\overset{\circ}{W}{}^2_2(G)$ with $(f,g) = \int \Delta f\, \Delta g\, dx$. It is easily seen that $f_G$ is the unique solution of the Dirichlet problem in this formulation.

However, the fact that $f - f_G \in \overset{\circ}{W}{}^2_2(G)$ does not in itself describe very clearly in what way $f_G$ and $\nabla f_G$ take the right boundary values, so a more explicit way of formulating the Dirichlet problem is desirable. This was undertaken by S. L. Sobolev, [31, §14], who formulated the Dirichlet problem in terms of the trace of f and $\nabla f$ on $\partial G$. Now, the most natural way of defining the trace is in terms of precisely defined functions, and this leads to the formulation of the "fine Dirichlet problem" given by Fuglede: (See [30].)

Let $f \in W_2^2(\mathbb{R}^d)$ be given. Find $u \in W_2^2(\mathbb{R}^d)$ such that $\Delta^2 u = 0$ in G, $u|_{G^c} = f|_{G^c}$, and $\nabla u|_{G^c} = \nabla f|_{G^c}$.

Note that $W_2^2(\mathbb{R}^d) \subset C(\mathbb{R}^d)$ if $d < 4$, and thus $u|_{G^c}$ is the ordinary restriction. Also the condition on $\nabla u$ is vacuous for parts of the boundary with (1,2)-capacity zero. This is for example the case for a (d-2)-manifold in $\mathbb{R}^d$.

This problem clearly has a solution, since $u = f_G$ satisfies the requirements. However, the uniqueness of the solution is no longer obvious. What has to be proved is that if $u \in D_2^2(G)$ and if $u|_{G^c} = 0$ and $\nabla u|_{G^c} = 0$, then $u \equiv 0$.

All this is easily generalized to e.g. the equation $\Delta^m u = 0$. The following is now immediate.

<u>Theorem</u> 3.3 (Fuglede [30]): The fine Dirichlet problem for $\Delta^m u = 0$ is uniquely solvable in the bounded open set G if and only if $G^c$ satisfies (m,2)-spectral synthesis.

In the case when $\partial G$ is a finite union of smooth manifolds (of arbitrary dimension) the uniqueness problem, and thus the spectral synthesis problem, was solved by Sobolev [31; §15]. Further results were given by Polking [28], and the author [22]. The following theorems are among those proved in [22]. (Theorem 3.4 (a) is in [28].)

Theorem 3.4:

(a) All closed $F \subset \mathbb{R}^d$ admit (m,p)-spectral synthesis for $m = 2,3,\ldots,$ if $p > d$.

(b) F admits (m,p)-spectral synthesis for $m = 2,3,\ldots,$ if $1 < p \leq d$, and

$$\sum_{n=1}^{\infty} \{2^{n(d-p)} C_{1,p}(F \cap B(x,2^{-n}))\}^{q-1} = \infty \text{ for (m,p)-q.e. } x \in F.$$

(Recall the Kellogg property, which says if $p > 2 - \frac{1}{d}$, the subset of F where the above infinite series converges has (1,p)-capacity zero.)

Theorem 3.5: A closed $F \subset \mathbb{R}^d$ with $C_{1,p}(F) = 0$ admits (m,p)-spectral synthesis for $m = 2,3,\ldots,$ provided either

(a) $2p > d$

or

(b) $$\liminf_{\delta \to 0} \frac{C_{2,p}(F \cap B(x,\delta))}{\delta^{d-2p}} > 0 \quad \text{for (m,p) q.e. } x \in F.$$

(A slight modification is needed if $d = 2p$.)

This theorem is not quite satisfactory. One would like to replace the lim inf in (b) by the divergence of a sum similar to the one in Theorem 3.4.

Using these two theorems and the Kellogg property it is not hard to prove the following.

Corollary 3.6: All closed $F \subset \mathbb{R}^d$ admit (m,p)-spectral synthesis, $m = 1,2,\ldots,$ if $p > \min(d/2, 2-1/d)$. In particular, the fine Dirichlet problem for $\Delta^m u = 0$ is uniquely solvable for all bounded G in $\mathbb{R}^2$ or $\mathbb{R}^3$.

The proof of Theorem 3.4 depends on the following estimate. (See [22].)

Lemma 3.7: Suppose that $f \in W_m^p(\mathbb{R}^d)$ and that $\nabla^k f|_F = 0$ for $0 \leq k \leq m-1$. Then for all balls $B(x,\delta)$ that intersect F

$$\int_{B(x,\delta)} |f(y)|^p \, dy \le A\delta^{mp} \int_{B(x,2\delta)} |\nabla^m f(y)|^p \, dy \quad \text{if} \quad p > d,$$

and

$$\int_{B(x,\delta)} |f(y)|^p \, dy \le A\delta^{mp} \frac{\delta^{d-p}}{C_{1,p}(F \cap B(x,\delta))} \int_{B(x,2\delta)} |\nabla^m f(y)|^p \, dy \text{ if } 1 < p \le d.$$

The proof of the theorem then consists in constructing a function $\omega$ in $C_0^\infty$ such that $0 \le \omega \le 1$, $\omega = 1$ on a neighborhood of F, (it is no restriction to assume that F is compact), and $\|\omega f\|_{m,p}$ is small. Then clearly $(1-\omega)f$ belongs to $\overset{\circ}{W}{}^p_m(F^c)$ and approximates f. $\omega$ has to be constructed in such a way that its derivatives match the factor $\delta^{d-p}/C_{1,p}(F \cap B(x,\delta))$ in the Lemma 3.7.

To show how this leads to the series condition, we prove the theorem under more restrictive conditions. (These are satisfied in most applications.).

Suppose that there is a sequence $\{a_n\}_1^\infty$, $a_n > 0$, such that for all $x \in F$

$$2^{n(d-p)} C_{1,p}(F \cap B(x,2^{-n})) \ge a_n,$$

and $\sum_{n=1}^\infty a_n^{q-1} = \infty$.

(Note that $a_n$ is bounded below if F satisfies for example a cone condition.)

We claim that F admits (m,p)-spectral synthesis.

Let $f \in W_m^p$, $\nabla^k f|_F = 0$, $0 \le k \le m-1$. For any integer n we let $G_n = \{x; \operatorname{dist}(x,F) < 2^{-n}\}$. Then one can easily construct a function $\omega_n \in C_0^\infty$ such that $0 \le \omega_n \le 1$, $\omega_n(x) = 1$ on $G_n$, $\omega_n(x) = 0$ on $G_{n-1}^c$, $|\nabla^k \omega_n(x)| \le A2^{kn}$, $1 \le k \le m$. We have to estimate $\int |\nabla^k \omega_n|^p |\nabla^{m-k} f|^p \, dx$ for $0 \le k \le m$. The case $k = 0$ is easily disposed of, since $\nabla^m f(x) = 0$ a.e. on F.

Now, applying the lemma to $\nabla^{m-k} f$ (which belongs to $W_k^p$), one easily sees that for any $x \in G_{n-1} \setminus G_n$

$$\int_{B(x,2^{-n})} |\nabla^{m-k} f|^p \, dy \le A2^{-kn} a_n^{-1} \int_{B(x,2^{-n+3})} |\nabla^m f|^p \, dy.$$

But $G_{n-1}\setminus G_n$ can be covered by balls $B(x_i, 2^{-n})$ in such a way that no point in $\mathbb{R}^d$ belongs to more than a fixed number $A = A_d$ of balls $B(x_i, 2^{-n+3})$. It follows that

$$\int_{G_{n-1}\setminus G_n} |\nabla^k \omega_n|^p \, |\nabla^{m-k} f|^p \, dy \le A\, a_n^{-1} \int_{G_{n-4}} |\nabla^m f|^p \, dy.$$

Since $\sum_1^\infty a_n^{q-1} = \infty$ and $a_n$ is bounded we can for arbitrarily large M find N > M so that $1 \le \sum_M^N a_n^{q-1} \le A$. We then set

$$\omega = \sum_M^N a_n^{q-1}\, \omega_n \Big/ \sum_M^N a_n^{q-1} .$$

Thus, $\omega \in C_0^\infty$, $\omega = 1$ on $G_N$, $\omega = 0$ on $G_{M-1}^c$, $0 \le \omega \le 1$, and $|\nabla^k \omega| \le a_n^{q-1} |\nabla^k \omega_n|$ on $G_n\setminus G_{n-1}$. Hence

$$\int_{\mathbb{R}^d} |\nabla^k \omega|^p |\nabla^{m-k} f|^p \, dx = \sum_M^N \int_{G_n\setminus G_{n-1}} |\nabla^k \omega|^p |\nabla^{m-k} f|^p \, dx \le$$

$$\le \sum_M^N a_n^q \int_{G_n\setminus G_{n-1}} |\nabla^k \omega_n|^p \, |\nabla^{m-k} f|^p \, dx$$

$$\le A \sum_M^N a_n^{q-1} \int_{G_{n-4}} |\nabla^m f|^p \, dx$$

$$\le A \sum_M^N a_n^{q-1} \int_{G_{M-4}} |\nabla^m f|^p dx \le A \int_{G_{M-4}} |\nabla^m f|^p dx.$$

This last integral is arbitrarily small, since $\int_F |\nabla^m f|^p \, dx = 0$. Thus $f \in \mathring{W}_m^p(F^c)$.

To prove Theorem 3.4 in general, one has to adapt the construction of $\omega$ to a situation where there is no uniformity in the divergence of the "Wiener series", which gets quite technical. See [22, Lemma 3.2]. Finally, in order to allow for an exceptional set $E \subset F$ where the Wiener series converges, one first has to adapt the function using the following lemma ([22], Lemma 5.2), which has some independent interest. (In fact, J. R. L. Webb has given an interesting application in the theory of non-linear PDE. See [34], where a self-contained proof of the lemma(without the set E), is also found.)

<u>Lemma</u> 3.8: Let $f \in W^p_m$, and let E be an arbitrary set with $C_{m,p}(E) = 0$. Then for any $\varepsilon > 0$ there exists a function $\omega \in W^p_m$ such that $0 \le \omega \le 1$, $\omega = 1$ on a neighborhood of E, $(1-\omega)f \in W^p_m \cap L^\infty$, $\|\omega f\|_{m,p} < \varepsilon$.

To prove Theorem 3.5, where $C_{1,p}(F) = 0$, Lemma 3.7 is of course useless. Also, the assumption $\nabla^{m-1}f|_F = 0$ is vacuous. The lemma is replaced by the following:

<u>Lemma</u> 3.9: Suppose that $f \in W^p_m$ and that $\nabla^k f|_F = 0$ for $0 \le k \le m-2$. Thenfor all balls $B(x,\delta)$ that intersect F

$$\int_{B(x,\delta)} |f|^p dy \le A\delta^{(m-1)p} \left\{ \int_{B(x,2\delta)} |\nabla^{m-1}f|^p dy + \delta^p \int_{B(x,2\delta)} |\nabla^m f|^p \, dy \right\}$$

if $2p > d$,

and

$$\int_{B(x,\delta)} |f|^p dy \le A\delta^{(m-1)p} \cdot \frac{\delta^{d-2p}}{C_{2,p}(F \cap B(x,\delta))} \left\{ \int_{B(x,2\delta)} |\nabla^{m-1}f|^p dy + \delta^p \int_{B(x,2\delta)} |\nabla^m f|^p dy \right\}$$

if $2p \le d$.

As before the theorem is proved by constructing a function $\omega$ such that $(1-\omega)f \in \mathring{W}^p_m(F^c)$ and approximates f. This time $\omega$ is constructed by means of "smooth truncation" of the capacitary potential for a neighborhood of f. The integrals $\int |\nabla^k \omega|^p |\nabla^{m-k} f|^p \, dx$ are estimated using the Whitney decomposition of $F^c$, a Harnack property for $\omega$, maximal functions, and an interpolation inequality for intermediate derivatives from [21].

The proofs of Theorem 3.4 and 3.5 can also be applied to open sets, and then they give sufficient conditions for stability, and consequently for $L^q$-approximation by solutions of $\Delta^m u = 0$. We have for example:

<u>Theorem</u> 3.10: A compact $K \subset \mathbb{R}^d$ is $(m,p)$-stable for all $p > d$, and for $p \leq d$ if

$$\sum_{n=1}^{\infty} \{2^{n(d-p)} C_{1,p}(B(x,2^{-n})\setminus K)\}^{q-1} = \infty$$

for $(m,p)$-q.e. $x \in \partial K$.

Different sufficient conditions are obtained by the argument used in proving Theorem 1.16.

<u>Theorem</u> 3.11: $K$ is $(m,p)$-stable if $(K^0)^c$ admits $(m,p)$-spectral synthesis, and $C_{k,p}(G\setminus K) = C_{k,p}(G\setminus K^0)$ for $k = 1,2,\ldots,m$.

<u>Theorem</u> 3.12: $K$ is $(m,p)$-stable if $(K^0)^c$ admits $(m,p)$-spectral synthesis and $K^c$ is $(k,p)$-thick $(k,p)$-q.e. on $\partial K$ for $k = 1,2,\ldots,m$.

In the converse direction, these sufficient conditions are not all necessary. In fact, if $K$ has no interior, one has the following necessary and sufficient condition.

<u>Theorem</u> 3.13: (Polking [28]). Let $K^0 = \emptyset$. Then

(a) $K$ is always $(m,p)$-stable if $mp > d$.

(b) $K$ is $(m,p)$-stable, $mp \leq d$, if and only if $C_{m,p}(G\setminus K) = C_{m,p}(G)$ for all open $G$.

Thus in order to show that the conditions in Theorem 3.11 are not necessary for, say $(2,2)$-stability in $\mathbb{R}^2$, all that has to be done is to produce a $K \subset \mathbb{R}^2$ such that $K^0 = \emptyset$ and $C_{1,2}(G\setminus K) < C_{1,2}(G)$ for some $G$, which is easy.

One can even ask if the condition in Theorem 3.13, $C_{m,p}(G\setminus K) = C_{m,p}(G\setminus K^0)$ for all open $G$, is not both necessary and sufficient for $(m,p)$-stability in the general case when $K$ has interior. In particular, are all $K \subset \mathbb{R}^d$ $(m,p)$-stable if $mp > d$, as in the case when $m = 1$?

The following example shows that this is not so if $p \leq d$. Thus, the presence of an interior really changes the situation. For simplicity we choose $m = p = d = 2$.

<u>Theorem</u> 3.14 ([22]): There is a compact $K \subset \mathbb{R}^2$ which is not (2,2)-stable. Thus the harmonic functions on K are not dense in $L_h^2(K)$.

<u>Proof</u>: We have to find a K and a $\varphi \in W_2^2$ such that $\varphi = 0$ on $K^c$ but $\nabla\varphi(x) \neq 0$ on a part of $\partial K$ with positive (1,2)-capacity. Let $B_0 = \{|x| < 1\}$, and let I be the interval $[-\frac{1}{2}, \frac{1}{2}]$ on the $x_1$-axis. Let $\{B_k\}_1^\infty$ be disjoint discs, $B_k = \{|x-x^{(k)}| < r_k\}$, with $x^{(k)} \in I$, and let $K = \bar{B}_0\backslash(\cup_1^\infty B_k)$.

Let $R_k > r_k$. Then one can easily construct a function $\chi_k \in C^\infty(\mathbb{R}^+)$ such that $\chi_k(r) = 1$ for $0 \le r \le r_k, \chi_k(r) = 0$ for $r \ge R_k$, $0 \le \chi_k \le 1$, $|\chi_k'(r)| \le \frac{A}{r}(\log R_k/r_k)^{-1}$, and $|\chi_k''(r)| \le A/r^2(\log R_k/r_k)^{-1}$.

Set $\psi_k(x) = \chi_k(|x-x^{(k)}|)$, and let $\varphi_0 \in C_0^\infty(B_0)$ be such that $\varphi_0(x) = x_2$ in a neighborhood of I. A short calculation shows that $\int|\nabla^2(\varphi_0\,\psi_k)|^2\,dx \le A(\log R_k/r_k)^{-1}$.

Now choose $\{R_k\}_1^\infty$, so that $\sum_1^\infty R_k \le \frac{1}{4}$, and choose $\{x^{(k)}\}_1^\infty$ so that the balls $\{|x-x^{(k)}| \le R_k\}$ are all disjoint, and dense on I. Finally choose $\{r_k\}_1^\infty$ so that $\sum(\log R_k/r_k)^{-1} < \infty$. Set $\varphi = \varphi_0(1-\sum_1^\infty \psi_k)$. Then clearly $\varphi \in W_2^2$ and $\varphi(x) = 0$ on $K^c$.

Every $x \in I\backslash(\cup_1^\infty\{|x-x^{(k)}| \le R_k)$ belongs to $\partial K$, and the set of such points has 1-dimensional Lebesgue measure $\ge \frac{1}{2}$, and thus positive (1,2)-capacity. But on any line perpendicular to I through such a point $\varphi(x) = \varphi_0(x) = x_2$ near I, and thus $\partial\varphi/\partial x_2 = 1$ on a subset of $\partial K$ of positive capacity. Thus $\varphi \notin \mathring{W}_2^2(K^0)$, which concludes the proof.

It thus appears that in order to give necessary and sufficient conditions for (m,p)-stability, a different way of measuring sets has to be found. Using a modified capacity È. M. Saak [29] has given such a condition in the case $p = 2$, $d > 2m$, but the situation is still not completely understood.

## REFERENCES

1. D. R. Adams, Maximal operators and capacity, Proc. Amer. Math. Soc. 34(1972), 152-156.

2. D. R. Adams and N. G. Meyers, Thinness and Wiener criteria for non-linear potentials, Indiana Univ. Math. J. 22(1972-73), 169-197.

3. D. R. Adams and N. G. Meyers, Bessel potentials. Inclusion relations among classes of exceptional sets, Indiana Univ. Math. J. 22(1972-73), 873-905.

4. I. Babuška, Stability of the domain with respect to the fundamental problems in the theory of partial differential equations, mainly in connection with the theory of elasticity, I, II. (Russian). Czechoslovak Math. J. 11(86)(1961), 76-105, and 165-203.

5. T. Bagby, Quasi topologies and rational approximation, J. Functional Analysis 10(1972), 259-268.

6. L. Bers, An approximation theorem, J. Analyse Math. 14(1965), 1-4

7. A. Beurling and J. Deny, Dirichlet spaces, Proc. National Acad. Sci. 45(1959), 208-215.

8. A. P. Calderón, Lebesgue spaces of differentiable functions and distributions, Proc. Symp. Pure Math. 4(1961), 33-49.

9. G. Choquet, Sur les points d'effilément d'un ensemble. Application à l'étude de la capacité. Ann. Inst. Fourier (Grenoble) 9 (1959), 91-101.

10. J. Deny, Systèmes totaux de fonctions harmoniques, Ann. Inst. Fourier (Grenoble) 1(1950), 103-113.

11. J. Deny, Méthodes hilbertiennes en théorie du potentiel, Potential Theory (CIME, I. Ciclo, Stresa 1969), 121-201, Ed. Cremonese, Rome 1970.

12. J. Deny and J.-L. Lions, Les espaces du type de Beppo Levi, Ann. Inst. Fourier (Grenoble) 5, 305-370 (1953-1954).

13. C. Fernström, On the instability of capacity, Ark. mat. 15(1977), 241-252.

14. B. Fuglede, Quasi topology and fine topology, Séminaire Brelot-Choquet-Deny, 10$^{e}$ année (1965-1966).

15. B. Fuglede, Applications du théoreme minimax à l'étude de diverses capacités, C. R. Acad. Sci. Paris, Sér. A266, 921-923 (1968).

16. A. A. Gončar, The property of "instability" of harmonic capacity, Dokl. Akad. Nauk SSSR 165(1965), 479-481. (Soviet Math. Dokl. 6(1965), 1458-1460).

17. V. P. Havin, Approximation in the mean by analytic functions. Dokl. Akad. Nauk SSSR 178 (1968), 1025-1028 (Soviet Math. Dokl. 9, (1968), 245-248).

18. V. P. Havin and V. G. Maz'ja, Nonlinear potential theory, Uspehi Mat. Nauk 27:6(1972), 67-138, (Russian Math. Surveys 27:6(1972), 71-148.

19. V. P. Havin and V. G. Maz'ja, Use of (p,ℓ)-capacity in problems of the theory of exceptional sets, Mat. Sb. 90 (132)(1973), 558-591. (Math. USSR-Sbornik 19 (1973), 547-580).

20. L. I. Hedberg, Non-linear potentials and approximation in the mean by analytic functions, Math. Z. 129 (1972), 299-319.

21. L. I. Hedberg, On certain convolution inequalities, Proc. Amer. Math. Soc. 36(1972), 505-510.

22. L. I. Hedberg, Two approximation problems in function spaces, Ark. mat. 16(1978), 51-81.

23. M. V. Keldyš, On the solubility and stability of the Dirichlet problem. Uspehi Mat. Nauk 8 (1941), 171-231. (Amer. Math. Soc. Translations (2) 51(1966), 1-73.)

24. N. S. Landkof, Foundations of modern potential theory, Nauka, Moscow 1966. (English translation, Springer-Verlag 1972).

25. Ju. A. Lysenko and B. M. Pisarevskii, Instability of harmonic capacity and approximations of continuous functions by harmonic functions, Mat. Sb. 76(118)(1968), 52-71. (Math. USSR Sbornik 5 (1968), 53-72).

26. M. S. Mel'nikov and S. O. Sinanjan, Problems in the theory of approximation of functions of one complex variable, Sovremennye problemy matematiki (ed. R. V. Gamkrelidze) 4 (1975), Itogi nauki i tehniki, VINITI, Moscow. (J. Soviet Math. 5(1976), 688-752).

27. N. G. Meyers, A theory of capacities for functions in Lebesgue classes, Math.Scand. 26(1970), 255-292.

28. J. Polking, Approximation in $L^p$ by solutions of elliptic partial differential equations, Amer. Math. J. 94(1972), 1231-1244.

29. È. M. Saak, A capacity criterion for a domain with stable Dirichlet problem for higher order elliptic equations, Mat. Sb. 100(142)(1976), 201-209. (Math. USSR-Sbornik 29(1976), 177-185.)

30. B. W. Schulze and G. Wildenhain, Methoden der Potentialtheorie für elliptische Differentialgleichugen beliebiger Ordnung, Akademie-Verlag, Berlin 1977.

31. S. L. Sobolev, Applications of functional analysis in mathematical physics, Izd. LGU, Leningrad, 1950 (English translation, Amer. Math. Soc., Providence, R.I., 1963).

32. E. M. Stein, Singular integrals and differentiability properties of functions, Princeton University Press, Princeton, N.J., 1970.

33. H. Wallin, Continuous functions and potential theory, Ark. mat. 5(1963), 55-84.

34. J. R. L. Webb, Boundary value problems for strongly nonlinear elliptic equations, J. London Math. Soc. (to appear).

FOURIER ANALYSIS OF MULTILINEAR CONVOLUTIONS,
CALDERÓN'S THEOREM, AND ANALYSIS ON LIPSCHITZ CURVES

R. R. Coifman
Washington University

and

Y. Meyer
Université de Paris-Sud, Centre d'Orsay

The purpose of this series of lectures is to describe certain methods of Fourier analysis useful in the study of operators which are not convolutions, or not even linear in some of their arguments. The main point of this presentation is to indicate that the Fourier transform remains a powerful tool in the analysis of these operators, much in the same spirit as in the study of $L^p$ multipliers $p \neq 2$.

In the first part we present a treatment of the first commutator by Fourier analysis and real variable methods. In the second half we introduce a Fourier transform for Lipschitz curves.

As a consequence we obtain a multiplier algebra of operators on such curves. These operators generalize the Cauchy integral studied by Calderón, and are based on Calderón's result. We present a proof of his theorem using some of the Fourier analysis developed for the first commutator.

## §1. The Calderón Commutator

Let $H$ denote the Hilbert transform,

$$Hf(x) = \lim_{\varepsilon\to 0} \frac{1}{\pi} \int_{|x-t|>\varepsilon} \frac{f(t)}{x-t}\, dt,$$

and let $Af = A(x)f(x)$ be the operator of multiplication by $A$. The Calderón commutator is the operator

$$C(f) = \lim_{\varepsilon\to 0} \frac{1}{\pi} \int_{|x-t|>\varepsilon} \frac{A(x)-A(t)}{(x-t)^2}\, f(t)dt .$$

This is the commutator between $H\frac{d}{dx}$ and $A$. A. P. Calderón [1] proved, by using complex variable methods (which will be explained later), that $C(f)$ is a bounded operator on $L^p$, $1 < p < \infty$, whenever $A' \in L^\infty$.

Like the Hilbert transform the Calderón commutator turns out to be a basic "building block" for various operators in analysis. Calderón was led to consider it in his precise calculus of pseudodifferential operators in higher dimensions. It also appears quite naturally in various other contexts, see [2], [3], [5].

Before studying this operator in some detail, various comments should be made:

a) The main difficulty in studying $C$ is in obtaining some estimate, say, in $L^2(\mathbb{R})$. Once such an estimate is obtained, one can extend it to other classes of functions by "real variable methods".

b) Any estimate on $L^2(\mathbb{R})$ of the form

$$\|C(f)\|_2 \leq N(A)\|f\|_2 ,$$

where $N(A)$ is a constant depending, say, on the $L^\infty$ norm of the first $10^{10}$ derivatives of $A$, can easily be shown (by dilating the functions) to be equivalent to

$$\|C(f)\|_2 \leq C_0\|A'\|_\infty\|f\|_2 .$$

We are thus forced to consider an estimate with $A' \in L^\infty$. (Observe that, by contrast, on the circle estimates involving $A''$ are trivial to obtain; the problem here is a global estimate on $\mathbb{R}$.)

c) The difficulty in obtaining an $L^2$ estimate for $A' \in L^\infty$ stems first from the fact that the kernel $\frac{A(x)-A(y)}{(x-y)^2}$ is not sufficiently "close" to $\frac{A'(y)}{x-y}$. Second, since the kernel is not a convolution kernel it is not clear how to use Fourier analysis.

d) It is convenient to think of $C(f)$ as a bilinear convolution operator. In fact, $C = C(a,f)$, $a = A'$, depends linearly on $a$ and $f$; moreover, it commutes with simultaneous translations of $a,f$. Formally this observation is "equivalent" to the possibility of writing

$$(1.1) \qquad C(a,f)(x) = \iint e^{ix(\xi+\alpha)}\sigma(\xi,\alpha)\hat{a}(\alpha)\hat{f}(\xi)d\alpha d\xi ,$$

where $\hat{f}$ is the Fourier transform of $f$ and $\sigma$ is a distribution. ($\sigma$ is a bilinear multiplier or the symbol of $C$ as a bilinear operator).

We are now ready to study $C$. The idea is to use a representation of the form (1.1) and to analyse $C$ by breaking $\sigma$ into simpler objects.

It is convenient to consider the commutator $C(a,f) - H(af) = [A,H]f'$ writing it as

$$[A,H]f'(x) = c\iint e^{ix(\xi+\alpha)}(\operatorname{sgn}(\xi+\alpha)-\operatorname{sgn}\xi)\frac{\xi}{\alpha}\hat{f}(\xi)\hat{a}(\alpha)d\xi d\alpha.$$

Here,

$$\sigma(\xi,\alpha) = \frac{\xi}{\alpha}(\operatorname{sgn}(\xi+\alpha)-\operatorname{sgn}\xi)$$

is quite easy to analyze. In fact, since $|\frac{\xi}{\alpha}| \le 1$ whenever $\sigma \neq 0$, we can represent this quantity as

$$|\frac{\xi}{\alpha}| = c\int_{-\infty}^{\infty} |\frac{\xi}{\alpha}|^{i\gamma} \frac{d\gamma}{1+\gamma^2}.$$

(We used the formula $e^{-s} = c\int_{-\infty}^{\infty} e^{-is\gamma}\frac{d\gamma}{1+\gamma^2}$, $s \ge 0$).

Combining the two identities we obtain

$$[A,H]f'(x) = c\iiint e^{ix(\xi+\alpha)}(\operatorname{sgn}(\xi+\alpha)-\operatorname{sgn}\xi)|\alpha|^{-i\gamma}\hat{a}(\alpha)|\xi|^{i\gamma}\hat{f}(\xi)\frac{d\xi d\alpha d\gamma}{1+\gamma^2}$$

$$= c\int_{-\infty}^{\infty}\frac{d\gamma}{1+\gamma^2}[a_{-\gamma},H]f_\gamma,$$

where $\hat{f}_\gamma = |\xi|^{i\gamma}\hat{f}$, $\hat{a}_{-\gamma} = |\alpha|^{-i\gamma}\hat{a}(\alpha)$, and

$$[a,H]f = a(Hf) - H(af) = \int_{-\infty}^{\infty}\frac{a(x)-a(y)}{x-y}f(y)dy.$$

At this stage we can make a few observations:

If $f \in L^2$ so is $f_\gamma$, with the same norm. Unfortunately if $a \in L^\infty$ $a_\gamma$ is in general unbounded and belongs to B.M.O. We thus are forced to study $[a,H]f$ for $a \in$ B.M.O. Another possibility is to assume, say, $f \in L^4$, $a \in L^4$. In this case $f_\gamma$, $a_\gamma$ are also in $L^4$ ($|\xi|^{i\gamma}$ being a Marcinkiewicz multiplier on $L^p$) and $[a_{-\gamma},H]f_\gamma$ is then in $L^2$, with $L^2$ norm not exceeding $c|\gamma|^\delta$ for some $0 < \delta < 1$. This yields an estimate of the form

$$\|[A,H\tfrac{d}{dx}]f\|_2 \le c\|a\|_4\|f\|_4$$

which by real variable techniques can be extended for other indices, see [4].

We now return to

$$[a,H]f = aHf - H(af)$$

with $f \in L^2$, $a \in$ B.M.O. This operator is in fact bounded on $L^2$ with

(1.2) $$\|[a,H]f\|_2 \le c\|a\|_{B.M.O.}\|f\|_2.$$

(1.2) can be proved in various ways, either directly by the so called "good $\lambda$ inequalities", or as a corollary of the weighted norm inequalities, see [5].

Perhaps the simplest proof of (1.2) uses the duality between $H^1$ and B.M.O. This approach permits us to write $a \in$ B.M.O. as

$$a = b_1 + H(b_2) \quad \text{with} \quad \|b_1\|_\infty + \|b_2\|_\infty \leq c\|a\|_{B.M.O.}$$

and so

$$[a,H]f = [b_1,H]f + [H(b_2),H]f.$$

We now employ the identity

$$Hb_2Hf - H(Hb_2f) = b_2f + H(Hfb_2)$$

to deduce the boundedness on $L^2$ of both terms.

Using the fact that $\|a_\gamma\|_{B.M.O.} \leq C_\varepsilon(|\gamma|+1)^{(1/2)+\varepsilon}, \forall\varepsilon > 0$, (see [6]) we obtain

Theorem I.

$$\|[A,H]f'\|_2 \leq c\|A'\|_{B.M.O.}\|f\|_2.$$

Calderón's equation,

$$h' = f'g, \tag{1.3}$$

is intimately related to the Calderón commutator and corresponds to the adjoint in $a$ of the commutator.
$f$ and $g$ are assumed to be given functions in $H^2(\mathbb{R})$ (i.e., $f,g \in L^2$ and $\hat{f}(x)$ and $\hat{g}(x)$ are $0$ for $x < 0$) and we look for $h$ in $H^1(\mathbb{R})$. Taking Fourier transforms we find that

$$\xi\hat{h}(\xi) = \int_{-\infty}^{\infty} \hat{g}(\xi-\eta)\eta\hat{f}(\eta)d\eta$$

or

$$\hat{h}(\xi) = \int_0^\xi \hat{g}(\xi-\eta)\left(\frac{\eta}{\xi}\right)\hat{f}(\eta)d\eta.$$

As before, this equals

$$\int_{-\infty}^{\infty}\left\{\int_0^\xi \hat{g}(\xi-\eta)\,\frac{\eta^{i\gamma}}{\xi^{i\gamma}}\,\hat{f}(\eta)d\eta\right\}\frac{d\gamma}{1+\gamma^2};$$

and consequently

$$h(x) = c \int_{-\infty}^{\infty} M_{-\gamma}(gM_\gamma f) \frac{d\gamma}{1+\gamma^2}$$

where $(M_\gamma f)^\wedge(\eta) = \eta^{i\gamma}\hat{f}(\eta)$.

Since $M_\gamma$ preserves $L^2$ and $H^1$ (see [4]) with norm $\leq c(1+|\gamma|)^{(1/2)+\varepsilon}$ we have the desired result. Observe that the equation does not in general have a solution in $L^1$ when f and g are merely in $L^2$.

We will return to this equation later using it's conformal invariance.

## §2. Carleson Measures, Commutators, and Multilinear Operators

The preceding method using both Fourier and Mellin analysis to study the commutator took into account the homogeneity of $\sigma$ as a function of $(\xi,\alpha)$. It is desirable to be able to extend our analysis to commutators of pseudodifferential operators, or even to general bilinear (multilinear) operators defined by reasonable symbols $\sigma$. In other words we would like to have the analog of a multiplier theorem for multilinear operators. To obtain such results we proceed as in the multiplier case by constructing certain basic operators which serve as building blocks (an equivalent formulation involves the Littlewood-Paley theory).

For simplicity we restrict our attention to the one dimensional bilinear case (see [5] for the general case.)

We let $\varphi,\psi$ be two test functions such that $\hat{\varphi},\hat{\psi}$ have compact support and $0 \notin \operatorname{supp} \hat{\psi}$.

We define

$$B(a,f) = \int_0^\infty (f*\varphi_t)(a*\psi_t) \frac{m(t)}{t}\, dt \tag{2.1}$$

where $\varphi_t(x) = \frac{1}{t}\varphi(x/t)$ and $m \in L^\infty(\mathbb{R})$.

We have

Theorem II.

$$\|B(a,f)\|_2 \leq c\|a\|_{B.M.O.}\|f\|_{L^2},$$

where $c$ depends on $\varphi,\psi,m$ alone.

We will see later that a superposition of the operators B (varying $\varphi,\psi,m$) yields various bilinear operators, and in particular the Calderón commutator.

To prove the theorem, we need a few simple facts concerning Carleson measures. Recall that a measure $\nu$ on $\mathbb{R}^2_+ = \{(x,t) \in \mathbb{R}^2 : t > 0\}$ is a Carleson measure if for each square $S = I \times (0,|I|)$ (here I is an interval and $|I|$ its measure) we have

$$\nu(S) \le c|I| \quad \text{with } c \text{ independent of } I.$$

A simple geometric argument shows that for any lower semi-continuous function $f(x,t)$ on $\mathbb{R}^2_+$

$$(2.2) \qquad \iint_{\mathbb{R}^2_+} |f(x,t)|^p \, d\nu \le C_p \int_{-\infty}^{\infty} N(f)^p(x)dx$$

where $N(f)(x) = \sup_{|x-y|<t} |f(y,t)|$ is the nontangential maximal function, see [8].

We also need the following lemmas.

<u>Lemma (2.3)</u>. Let $\psi$ be defined as above. If $f \in L^2$ then

$$\int_{-\infty}^{\infty}\int_0^{\infty} |f*\psi_t|^2 \frac{dt}{t} \, dx = c \int_{-\infty}^{\infty} |f(x)|^2 \, dx.$$

This is immediate by Plancherel's theorem.

<u>Lemma (2.4)</u>. With $\psi$ as above and $a \in$ B.M.O. ,

$$d\nu(x,t) = |\psi_t * a(x)|^2 \frac{dxdt}{t}$$

is a Carleson measure with

$$\nu(S) \le c\|a\|^2_{B.M.O.}|I| \quad \text{and } c \text{ depending only on } \psi,$$

see [6].

This is an immediate consequence of (2.3) combined with simple geometric arguments.

<u>Proof of Theorem II</u>. We start by writing $\varphi = \varphi_0 + \psi_0$ where $0 \in \operatorname{supp} \hat{\varphi}_0$ but $0 \notin \operatorname{supp} \hat{\varphi}_0 + \operatorname{supp} \hat{\psi}$ and $0 \notin \operatorname{supp} \hat{\psi}_0$ (just choose the support of $\hat{\varphi}_0$ to be sufficiently small); here we can take $\hat{\varphi}_0$, $\hat{\psi}_0$ in $C_0^\infty(\mathbb{R})$.

Correspondingly we have

$$B(a,f) = B_0(a,f) + B_1(a,f),$$

where

$$B_0(a,f) = \int_0^\infty (f*\varphi_{0,t})(a*\psi_t)\, \frac{m(t)dt}{t}$$

and

$$B_1(a,f) = \int_0^\infty (f*\psi_{0,t})(a*\psi_t)\, \frac{m(t)}{t}\, dt.$$

To study $B_0$ we introduce an even test function $\psi_1$ whose Fourier transform $\hat{\psi}_1$ is 1 on supp $\hat{\varphi}_0$+ supp $\hat{\psi}$ and such that $0 \notin \text{supp}\, \hat{\psi}_1$. Clearly

$$\psi_{1,t}*[(f*\varphi_{0,t})(a*\psi_t)] = (f*\varphi_{0,t})(a*\psi_t),$$

since $\hat{\psi}_{1,t} \equiv 1$ on $\text{supp}((f*\varphi_{0,t})(a*\psi_t))^\wedge$.

We can rewrite $B_0(a,f)$ as

$$B_0(a,f) = \int_0^\infty \psi_{1,t}*[(f*\varphi_{0,t})(a*\psi_t)]\, \frac{m(t)}{t}\, dt.$$

To estimate the $L^2$ norm of this term it suffices to estimate

$$\int_{-\infty}^\infty gB_0(a,f)dx = I \quad \text{for} \quad g \in L^2.$$

We have

$$I = \int_{-\infty}^\infty\int_0^\infty (g*\psi_{1,t}(x))(f*\varphi_{0,t}(x)(a*\psi_t(x))\, \frac{m(t)dxdt}{t}$$

$$\leq \left(\int_{-\infty}^{-\infty}\int_0^\infty |g*\psi_{1,t}(x)|^2\, \frac{dxdt}{t}\right)^{1/2} \left(\int_{-\infty}^\infty\int_0^\infty |f*\varphi_{0,t}(x)|^2 |a*\psi_t(x)|^2\, \frac{dxdt}{t}\right)^{1/2}$$

$$\leq \|g\|_2\left(\int_{-\infty}^\infty |f^*(x)|^2 dx\right)^{1/2}\|a\|_{B.M.O.} \leq \|a\|_{B.M.O.}\|g\|_2\|f\|_2,$$

where the first integral is estimated by Lemma (2.3), while the second integral is estimated by Lemma (2.4) and (2.2) using the fact that $N(f*\varphi_{0,t}) \leq cf^*(x)$. (Here, $f^*(x)$ is the Hardy-Littlewood maximal function, and $|a*\psi_t|^2\, \frac{dxdt}{t}$ is a Carleson measure).

The estimate for $B_1$ is similar. We choose $\varphi_1$ even, such that $\hat{\varphi}_1 \equiv 1$ on supp $\hat{\psi}_0$+ supp $\hat{\psi}$ and rewrite $B_1(a,f)$ as

$$B_1(a,f) = \int \varphi_{1,t}*((f*\psi_{0,t})(a*\psi_t))\, \frac{m(t)dt}{t}\,.$$

We again estimate $I = \int gB_1(a,f)dx$

$$= \iint (g*\varphi_{1,t})(f*\psi_{0,t})(a*\psi_t) \frac{dxm(t)dt}{t}$$

$$\leq \left(\int |g*\varphi_{1,t}|^2 |a*\psi_t|^2 \frac{dxdt}{t}\right)^{1/2} \left(\int |f*\psi_{0,t}|^2 \frac{dxdt}{t}\right)^{1/2}$$

as before by interchanging the roles of $f$ and $g$.

A variation on this theorem involves

$$\overline{B}(a,f) = \int_0^\infty (f*\psi_t)(a*\varphi_t)m(t) \frac{dt}{t}$$

where $a$ is assumed to be bounded.

We have the following estimate:

(2.5) $$\|\overline{B}(a,f)\|_2 \leq c\|a\|_\infty \|f\|_2 .$$

This estimate is obtained as before. We write $\varphi = \varphi_0 + \psi_0$ with $0 \notin \operatorname{supp} \hat{\varphi}_0 + \operatorname{supp} \hat{\psi}$. The technique of Theorem II is used to deal with

$$\int_0^\infty (f*\psi_t)(a*\psi_{0,t})m(t) \frac{dt}{t} .$$

The term

$$\int_0^\infty (f*\psi_t)(a*\varphi_{0,t})m(t) \frac{dt}{t}$$

can be rewritten as

$$I(x) = \int_0^\infty \psi_t' * [(f*\psi_t)(a*\varphi_{0,t})] \frac{dt}{t}$$

where $\hat{\psi}'$ is 1 on $\operatorname{supp} \hat{\varphi}_0 + \operatorname{supp} \hat{\psi}$ and $0 \notin \operatorname{supp} \hat{\psi}'$.

For $g \in L^2$ we make the estimate,

$$\int g(x)I(x)dx = \iint_0^\infty (g*\psi_t')\cdot(f*\psi_t)(a*\varphi_{0,t})m(t) \frac{dt}{t}$$

$$\leq \int dx \left(\int_0^\infty |g*\psi_t'|^2 \frac{dt}{t}\right)^{1/2} \left(\int_0^\infty |f*\psi_t|^2 \frac{dt}{t}\right)^{1/2} \|a\|_\infty \leq c\|f\|_2 \|g\|_2 \|a\|_\infty ,$$

and (2.5) follows.

As an application we prove the following generalization of the Calderón commutator estimate.

Theorem III. Let $m(\xi)$ satisfy $|m^{(k)}(\xi)| \leq C_k |\xi|^{-k}$, $k = 0,1,2,\ldots$ . Define the operator $M$ by

$$Mf = (m\hat{f})^{\vee}.$$

Then the commutator between $M$ and multiplication by $A$ is smoothing of order $1$, i.e.,

$$\|[M,A]f'\|_2 \leq c\|A'\|_\infty \|f\|_2.$$

Proof. We write as before

$$[M,A](f')(x) = c \iint e^{ix(\xi+\alpha)}(m(\xi+\alpha) - m(\xi)) \frac{\xi}{\alpha} \hat{f}(\xi)\hat{a}(\alpha)d\xi d\alpha$$

and introduce a "partition" of unity permitting the comparison of the relative size of $\xi$ and $\alpha$.

First we write

$$1 = \hat{\varphi}_0(\tfrac{\xi}{\alpha}) + \hat{\psi}_0(\tfrac{\xi}{\alpha}),$$

where $\hat{\varphi}_0(t)$ is supported in $|t|<10$ and $\hat{\psi}_0(t)$ in $|t|>5$, and then split the corresponding integrals into $I + J$ where

$$I = \iint e^{ix(\xi+\alpha)}m(\xi+\alpha)\hat{\varphi}_0(\tfrac{\xi}{\alpha}) \frac{\xi}{\alpha} \hat{f}(\xi)\hat{a}(\alpha)d\xi d\alpha$$

$$- \iint e^{ix(\xi+\alpha)}m(\xi)\hat{\varphi}_0(\tfrac{\xi}{\alpha}) \frac{\xi}{\alpha} \hat{f}(\xi)\hat{a}(\alpha)d\xi d\alpha$$

$$= I_1 - I_2$$

and

$$J = \iint e^{ix(\xi+\alpha)}(m(\xi+\alpha) - m(\xi)) \frac{\xi}{\alpha} \hat{\psi}_0(\tfrac{\xi}{\alpha})\hat{f}(\xi)\hat{a}(\alpha)d\xi d\alpha.$$

To study $I_1$ (or $I_2$) we choose $\hat{\psi}$ even in $C_0^\infty(\mathbb{R})$, supported in $1<|t|<3$, and such that

$$1 = \int_0^\infty \hat{\psi}^2(t) \frac{dt}{t} = \int_0^\infty \hat{\psi}^2(\alpha t) \frac{dt}{t}.$$

Clearly,
$$\hat{\varphi}_0(\tfrac{\xi}{\alpha})\hat{\psi}^2(\alpha t) = \hat{\varphi}_0(\tfrac{\xi}{\alpha})\hat{\psi}^2(\alpha t)\hat{\varphi}_1^2(\xi t)$$

for any $\hat{\varphi}_1(t)$ in $C_0^\infty(\mathbb{R})$ which is equal to $1$ for $|t|<30$ (since $|\xi|<10|\alpha|$ and $|\alpha|<\frac{3}{t}$). This permits us to rewrite $I_1$ as

$$I_1 = \int_0^\infty \iint e^{ix(\xi+\alpha)}m(\xi+\alpha)\hat{\varphi}_0(\tfrac{\xi}{\alpha}) \frac{\xi}{\alpha} \hat{\psi}^2(\alpha t)\hat{\varphi}_1^2(\xi t)\hat{f}(\xi)\hat{a}(\alpha)d\xi d\alpha \frac{dt}{t}.$$

We now observe that the function

$$\hat{\varphi}_0(\tfrac{s}{r}) \frac{s}{r} \hat{\psi}(r)\hat{\varphi}(s)$$

is in $C_0^\infty(\mathbb{R}^2)$ and can be represented as

$$\iint e^{i(su+rv)}\eta(u,v)\,dudv$$

(where $\eta$ is its Fourier transform). Taking $s = \xi t$, $r = \alpha t$ and introducing this in the expression for $I_1$ we get

$$I_1 = \int \eta(u,v)\,dudv \int_0^\infty\iint e^{ix(\xi+\alpha)}m(\xi+\alpha)\hat{\psi}(\alpha t)\hat{\varphi}_1(\xi t)e^{i(\alpha tv+\xi tu)}\hat{f}(\xi)\hat{a}(\alpha)\,\frac{d\alpha d\xi dt}{t}$$

$$= \int_{\mathbb{R}^2} \eta(u,v)M(B_{u,v}(a,f))\,dudv ,$$

where $Mf = (m\hat{f})^{\vee}$ and

$$B_{u,v}(a,f) = \int_0^\infty f*\varphi_t^u\ a*\psi_t^v\ \frac{dt}{t}$$

with $\hat{\varphi}^u(\xi) = \hat{\varphi}_1(\xi)e^{i\xi u}$, $\hat{\psi}^v(\alpha) = \hat{\psi}(\alpha)e^{i\alpha v}$.

Using Theorem II and the rapid decay of $\eta$ we obtain the desired estimate for $I_1$ and $I_2$ ($I_2$ is studied by the same argument).

As for $J$ we take $\hat{\psi}$ as before and write

$$1 = \int_0^\infty \hat{\psi}^2(\xi t)\,\frac{dt}{t} .$$

We again have

$$\hat{\psi}_0(\tfrac{\xi}{\alpha})\hat{\psi}^2(\xi t) = \hat{\psi}_0(\tfrac{\xi}{\alpha})\hat{\psi}^2(\xi t)\hat{\varphi}^2(\alpha t)$$

where $\hat{\varphi}(u)$ is 1 for $|u|<30$. We get

$$J = \int_0^\infty\iint e^{ix(\xi+\alpha)}\sigma(\xi,\alpha,t)\hat{\psi}(\xi t)\hat{\varphi}(\alpha t)\hat{f}(\xi)\hat{a}(\alpha)\,d\alpha d\xi\,\frac{dt}{t}$$

with $\sigma(\xi,\alpha,t) = (m(\xi+\alpha) - m(\xi))\,\frac{\xi}{\alpha}\,\hat{\psi}_0(\frac{\xi}{\alpha})\hat{\psi}(\xi t)\hat{\varphi}(\alpha t)$.

Observe that $\left|\frac{\partial^{k+j}}{\partial s^k\partial r^j}\,\sigma(\frac{s}{t},\frac{r}{t},t)\right| \le C_{k,j}$

uniformly in $t$, and that this expression has support in $|s|+|r|\le 40$. Thus using the Fourier inversion formula we can write

$$\sigma(\tfrac{s}{t},\tfrac{r}{t},t) = \iint_{\mathbb{R}^2} e^{i(su+rv)}\eta(t,u,v)\,dudv$$

with $\eta(t,u,v) \le \frac{c}{(1+|u|^2+|v|^2)^N}$. Substituting in this formula by $s = \xi t$, $r = \alpha t$ we get

$$J = \iint_{\mathbb{R}^2} \frac{dudv}{(1+|u|^2+|v|^2)^N} \bar{B}_{u,v}(a,f)$$

where

$$\bar{B}_{u,v}(a,f) = \int_0^\infty (f*\psi_t^u)(a*\varphi_t^v)\eta(t,u,v)(1+|u|^2+|v|^2)^N \frac{dt}{t}$$

is estimated using (2.5).

We end this section by quoting a general bilinear multiplier theorem on $\mathbb{R}^n$ which can easily be reduced to Theorem III.

<u>Theorem IV</u>. Let $\sigma(D)(a,f) = \int_{\mathbb{R}^n}\int_{\mathbb{R}^n} e^{ix(\xi+\alpha)}\sigma(\alpha,\xi)\hat{f}(\xi)\hat{a}(\alpha)d\xi d\alpha$

where

$$|\partial_\alpha^p \partial_\xi^q \sigma(\alpha,\xi)| \le c_{p,q}(|\alpha|+|\xi|)^{-|p|-|q|}, (\alpha,\xi) \ne (0,0);$$

then

$$\|\sigma(D)(a,f)\|_2 \le c\|a\|_\infty \|f\|_2.$$

Moreover, if $\sigma(0,\xi) \equiv 0$ then

$$\|\sigma(D)(a,f)\|_2 \le c\|a\|_{B.M.O.} \|f\|_2,$$

see [5].

## §3. <u>Fourier Analysis on Lipschitz Curves and Calderón's Theorem on the Cauchy Integral</u>

We now would like to introduce the notion of a Fourier transform on Lipschitz curves. As a consequence we will obtain a new algebra of pseudodifferential operators on such curves. We will also sketch a proof of Calderón's theorem indicating the connection with §1.

As in Calderón's paper we study curves $\Gamma = \{y=\varphi(x)\}$ where $\|\varphi'\|_\infty \le \delta_0$ ($\delta_0$ is some small constant). For simplicity we assume first that $\varphi' \in C_0^\infty(\mathbb{R})$ (our final estimates will depend only on $\|\varphi'\|_\infty$). We start by introducing a class $A(\Gamma)$ of functions which is dense in $L^2(\Gamma)$. ($L^2(\Gamma)$ is identified with $L^2(\mathbb{R})$).

<u>Definition</u>. $A(\Gamma) = \{f(z)$ holomorphic for $\min\varphi - \varepsilon < \mathrm{Im}\, z < \max\varphi +$ and such that

$$\int |f(x+iy)|^2\, dx < c_\varepsilon$$

in this region}, i.e., $A(\Gamma)$ is the space of functions holomorphic in some open horizontal strip containing $\Gamma$, which are in $L^2$ on each horizontal line.

Lemma (3.1). $A(\Gamma)$ is dense in $L^2(\Gamma)$.

In fact, assuming that g is orthogonal to $A(\Gamma)$, we have

$$0 = \int g_0(z) \frac{dz}{z-z_0}, \quad g_0 = g(z) \frac{1}{z'} \quad \text{and } |\mathrm{Im}\, z_0| > \|\varphi\|_\infty .$$

By analytic continuation this remains valid for all $z_0 \notin \Gamma$. Finally, for almost all $z_0 \in \Gamma$, we have

$$g_0(z_0) = \lim_{\delta \to 0} \frac{1}{2\pi i} \left\{ \int_\Gamma g_0(z) \left( \frac{1}{z-z_0-i\delta} - \frac{1}{z-z_0+i\delta} \right) dz \right\};$$

as a consequence, $g_0 \equiv 0$.

For a function f in $A(\Gamma)$ we define the Fourier transform $\hat{f}$ as

$$(3.2) \qquad \hat{f}(t) = \int_\Gamma e^{-itz} f(z)dz \quad \left( = \int_{-\infty}^{\infty} e^{-itx} f(x)dx \right).$$

It is easy to check that

$$(3.3) \qquad f(z) = \frac{1}{2\pi} \int_{-\infty}^{\infty} e^{itz} \hat{f}(t)dt$$

(this relation is actually valid on the strip containing $\Gamma$).

Another important relation is

Lemma (3.4). If $f, g \in A(\Gamma)$ then

$$\int_\Gamma f(z)g(z)dz = \frac{1}{2\pi} \int_{-\infty}^{\infty} \hat{f}(t)\hat{g}(-t)dt.$$

The proof is trivial, since by Cauchy's theorem we can deform $\Gamma$ to $\mathbb{R}$. The formula then follows from Plancherel's theorem.

The main defect of this formula is the absence of an $L^2$ estimate for the Fourier transform.

We should also mention that an alternate way of introducing the Fourier transform is by formula (3.2) for, say, $L^1$ functions with compact support. Then one has to interpret the inversion formula (3.3) as

$$f(z) = \lim_{\varepsilon \to 0} \frac{1}{2\pi} \int_{-\infty}^{\infty} e^{-\varepsilon|t|^2} e^{itz} \hat{f}(t)dt$$

(which can be shown to exist for $z \in \Gamma$).

Our purpose now is to study operators of the form

$$M(f)(z) = \frac{1}{2\pi}\int_{-\infty}^{\infty} e^{itz} m(t)\hat{f}(t)dt \ , \ z \in \Gamma$$

on $L^2(\Gamma)$. (The case $p \neq 2$ will be an immediate consequence). We have

Theorem V. $M$ is bounded on $L^2(\Gamma)$ if $m(t)$ admits a bounded holomorphic extension to the domain

$$|\text{Im } z| < \beta|\text{Re } z| \text{ for some } \beta \leq \delta_0$$

where the Calderón condition, $\|\varphi'\|_\infty < \delta_0$, is satisfied.

Actually the condition on $m$ is necessary if we require that $M$ be bounded for all curves with a given Lipschitz constant.

The simplest multiplier verifying this hypothesis is $m(\xi) = \begin{cases} 1 & \xi > 0 \\ 0 & \xi \leq 0 \end{cases}$. A simple application of Theorem V gives

$$Mf(z) = \frac{1}{2\pi i}\lim_{\varepsilon\to 0}\int_\Gamma \frac{f(\zeta)}{\zeta-(z+i\varepsilon)}d\zeta \ .$$

The inequality in $L^2$ for this operator is Calderón's theorem for the Cauchy integral on Lipschitz curves [3]. This inequality is known to be valid if $\|\varphi'\|_\infty \leq \delta_0$ for some small $\delta_0 > 0$. It is an open problem for a general Lipschitz curve. The theorem stated above will be seen to be a consequence of Calderón's theorem, as well as certain results of C. Kenig concerning Hardy Spaces on Lipschitz domains.

We now sketch a proof of Calderón's theorem. Except for some minor changes (designed to relate it to the analysis of §1) the proof follows Calderón's paper [3].

The idea is to view the Cauchy integral as a perturbation of the Hilbert transform. Letting $z_\lambda(x) = x + i\lambda\varphi(x)$, where $\varphi'(x) = a(x)$ is in $L^\infty$, we study the variation of the operators

$$C_\lambda(f)(x) = \frac{-1}{2\pi i}\lim_{\varepsilon\to 0}\int \frac{f(t)dz_\lambda(t)}{z_\lambda(x)-z_\lambda(t)-i\varepsilon}.$$

(For simplicity here, $\varphi \in C_0^\infty$, and we are only interested in an à priori estimate depending on $\|\varphi'\|_\infty$ alone).

Calderón's idea is to show that

(3.5) $$\frac{\partial}{\partial\lambda}\|C_\lambda\| \le c\|C_\lambda\|^2$$

where $\|C_\lambda\|$ is the $L^2$ operator norm of $C_\lambda$ and $c$ is a fixed universal constant.

Integrating this inequality one obtains

$$\|C_\lambda\| \le \frac{\|C_0\|}{1-\lambda c\|C_0\|}$$

which is bounded as long as $\lambda \le \delta_0$ is sufficiently small. This gives the theorem for sufficiently small Lipschitz constants.

To obtain (3.5) we calculate

$$\frac{\partial}{\partial\lambda} C_\lambda = c \lim_{\varepsilon\to 0} \int \frac{\varphi(x)-\varphi(t)}{(z_\lambda(x)-z_\lambda(t)-i\varepsilon)^2} f(t)dz_\lambda(t) + C_\lambda(f_0)$$

where $f_0 = if(t)\,\frac{\varphi'(t)}{1+i\lambda\varphi(t)}$.

To estimate the first term, which looks formally like the Calderón commutator, we can use the "curve Fourier" transform to rewrite it as

$$F(z) = c \iint e^{iz(\alpha+t)}\,\frac{t\chi(t)-(t+\alpha)\chi(t+\alpha)}{\alpha}\,\hat{f}(t)\hat{a}(\alpha)dtd\alpha$$

where

$$\hat{a}(\alpha) = \int e^{-i\alpha z(x)}\varphi'(x)dz(x) \quad \text{and} \quad \chi(t) = \begin{cases} 1 & t > 0 \\ 0 & t < 0. \end{cases}$$

We need to estimate

$$I = \int_{\Gamma_\lambda} g(z)F(z)dz \quad \text{for } g \in L^2(\Gamma_\lambda).$$

Using (3.4) we obtain

$$I = \int_\Gamma G(z)\varphi'(z)dz$$

where $\hat{G}(\alpha) = c\int \hat{g}(\alpha-t)\hat{f}(t)\,\frac{t\chi(t)-(t-\alpha)\chi(t-\alpha)}{\alpha}\,dt$.

We now write $f = f_+ + f_-$, $g = g_+ + g_-$ where

$$f_+ = C_\lambda(f) \text{ has Fourier transform } \hat{f}\chi.$$

We have two trivial cases corresponding to $g_-, f_+$ and $g_+, f_-$. In the first the integral reduces to $g_-(z)f_+(z)$. In the second the integral is 0. The case $g_+, f_+$ gives

$$\hat{h}(\alpha) = \int \hat{g}_+(\alpha-t)\hat{f}_+(t)\frac{t}{\alpha}\,dt,$$

that is,

$$h(z) = \int^z f_+'(u)g_+(u)du.$$

This is Calderón's equation for functions holomorphic above $\Gamma$. The case $-,-$ is similar.

The function $h$ above is easily estimated in terms of $f_+, g_+$. We just observe that the equation is invariant under changes of variables, i.e., if

$$h' = f_+'g_+$$

and

$$H = h\circ\emptyset, \quad F = f_+\circ\emptyset, \quad G = g_+\circ\emptyset,$$

then $H' = F'G$.

If we let $\emptyset$ be the conformal map from the upper half plane onto the region above $\Gamma$ we find the equation studied in §1. However the estimate we need is

$$\int |H(x)||\emptyset'(x)|dx \le c\left(\int |F(x)|^2|\emptyset'(x)|dx\right)^{1/2}\left(\int |G(x)|^2|\emptyset'(x)|dx\right)^{1/2}.$$

This follows easily from the fact that $|\emptyset'(x)|$ is an $A_2$ weight (see [ 7 ]) and our treatment of the equation. Altogether we find

$$|I| \le \sum \|f_\pm\|_2\|g_\pm\|_2 \le c\|C_\lambda\|^2\|f\|_2\|g\|_2.$$

Consequently

$$\frac{\partial}{\partial\lambda}\|C_\lambda\| \le \left\|\frac{\partial}{\partial\lambda}C_\lambda\right\| \le c(\|C_\lambda\|^2 + \|C_\lambda\|) < c\|C_\lambda\|^2$$

and so (3.5), and, hence, Calderón's theorem, is proved.

We are now ready to prove Theorem V. We write $\theta_0 = \tan\beta$, $0 < \theta_0 < \frac{\pi}{2}$, and claim that the operator $M$ can be realized as $M_+ + M_-$ where $M_+$ corresponds to the restriction of $m$ to $t > 0$ and

$$(3.6) \qquad M_+f(z) = \lim_{\delta\downarrow 0}\int_\Gamma k(z+i\delta-\zeta)f(\zeta)d\zeta$$

where

$$k(z) = \int_0^\infty e^{itz}m(t)dt.$$

The integral for $k(z)$ is defined for $\operatorname{Re} z > 0$, and, by analytic continuation elsewhere (see below); also $k$ is holomorphic in

$$z = Re^{i\varphi} \quad -\theta_0 < \varphi < \theta_0 + \pi$$

and satisfies

$$|k(z)| \le \frac{C_\varepsilon}{|z|} \quad \text{for} \quad -\theta_0 + \varepsilon < \varphi < \theta_0 + \pi - \varepsilon_0 .$$

In fact if we let $k_\theta(\zeta) = e^{i\theta}\int_0^\infty e^{it\zeta}m(e^{i\theta}t)dt$, which is defined for $|\theta| < \theta_0$ and $\mathrm{Re}\zeta > 0$, we find by an obvious change in contour that

$$k(z) = k_\theta(e^{i\theta}z).$$

Since $k_\theta(\zeta)$ is analytic in $\mathrm{Re}\zeta > 0$ we find the analytic continuation of $k$ whenever $\mathrm{Re}\, e^{i\theta}z > 0$ or $-\theta_0 < \arg z < \theta_0 + \pi$.

Finally a familiar integration by parts argument gives

$$|k^{(j)}(z)| = O(|z|^{-j-1}), \varepsilon_0 - \theta_0 < \arg z < \theta_0 + \pi - \varepsilon_0 .$$

Formula (3.6) follows from Lemma (3.4) whenever $\mathrm{Im}\, z > \|\varphi\|_\infty$, and by analytic continuation elsewhere.

Observe also that by Cauchy's theorem, using the holomorphy and decay of $k$, we have

$$M_+f = M_+f_+ , \quad f_+ = C(f).$$

Before proving the $L^2$ estimate for $M_+$ ($M_-$ admits the same treatment) we should point out that the method described below is an adaptation of the Littlewood-Paley theory to our case, see Stein [8].

The basic tools needed are a few equivalent norms for the Hardy space $H^2$ of $L^2$ holomorphic functions above $\Gamma$. These were proved by C. Kenig in his thesis [7] and by B. Dahlberg in $\mathbb{R}^n$.

Lemma (3.7). If $h$ is holomorphic above $\Gamma$ the following norms are equivalent:

$1^\circ$ $\quad \sup_{\eta>0} \left(\int |h(x+i\varphi(x)+i\eta)|^2 dx\right)^{1/2}$

$2^\circ$ $\quad \left(\int_{-\infty}^{\infty}\int_0^\infty \eta^{2j-1}|h^{(j)}(x+i\varphi(x)+i\eta)|^2 dx d\eta\right)^{1/2} \qquad j = 1,2,\ldots .$

Given $z_0 \in \Gamma$ we define $z = z_0 + i\eta, \eta > 0$, $z^* = z_0 + i\frac{\eta}{2}$, and $\Gamma^* = \Gamma + i\frac{\eta}{2}$. We need the following simple geometric observation.

Lemma (3.8). If $\|\varphi'\|_\infty \le \delta$ then $\exists C_\delta$ s.t. for all $\zeta \in \Gamma^*$, $|\zeta - z| > C_\delta[(x-x_0)^2+\eta^2]^{1/2}$.

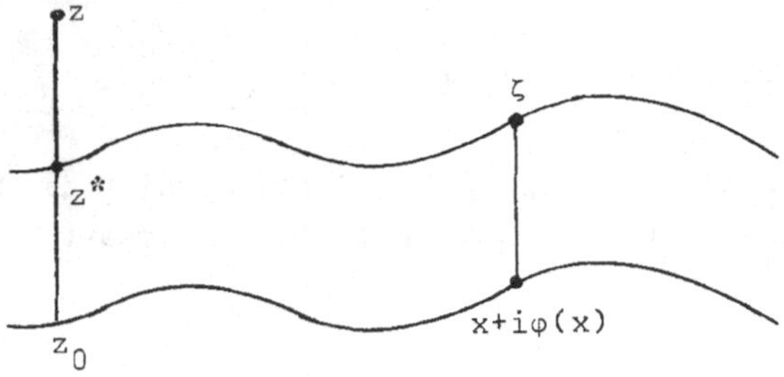


$$\zeta = x + i\varphi(x) + i\,\frac{\eta}{2}$$

To estimate the $L^2$ norm of

$$h(z) = \int_\Gamma k(z-\zeta)f_+(\zeta)d\zeta \;, \operatorname{Im} z > \varphi(\operatorname{Re} z)$$

we estimate $h''(z)$ and then use Lemma (3.7). We have by integration by parts and change of contour,

$$h''(z) = \int_\Gamma k''(z-\zeta)f_+(\zeta)d\zeta = \int_\Gamma k'(z-\zeta)f_+'(\zeta)d\zeta = \int_{\Gamma^*} k'(z-\zeta)f_+'(\zeta)d\zeta.$$

Using our estimate on $k'$ we get

$$|h''(z)| \le C \int_{\Gamma^*} \frac{|f_+'(\zeta)|}{|z-\zeta|^2}\, ds$$

$$\le C\left(\int_{\Gamma^*} \frac{ds}{|z-\zeta|^2}\right)^{1/2}\left(\int_{\Gamma^*} \frac{|f_+'(\zeta)|^2}{|z-\zeta|^2}\, ds\right)^{1/2}$$

$$\le C\;\eta^{-1/2}\left(\int_{\Gamma^*} \frac{|f_+(\zeta)|^2}{|z-\zeta|^2}\, ds\right)^{1/2}.$$

Therefore

$$\int_{-\infty}^{\infty}\int_0^{\infty} |h''(x+i\varphi(x)+i\eta)|^2 \eta^3 dx d\eta \le C\int_0^{\infty}\iint \eta^2\, \frac{|f_+'(t+i\varphi(t)+i\,\frac{\eta}{2})|^2}{(t-x)^2 + \eta^2}\, dt dx d\eta$$

$$= C' \int_0^{\infty}\int \eta |f_+'(t+i\varphi(t)+i\eta)|^2 dt d\eta.$$

Using Lemma (3.7) we find

$$\|M_+(f)\|_2 \le C\|f_+\|_2 \;.$$

(This result is valid without the restriction $\|\varphi'\|_\infty \le \delta_0$).

Using Calderón's theorem, assuming $\|\varphi'\|_\infty \le \delta_0$, we get

$$\|M_+(f)\| \le C\|f\|_2,$$

and this completes the proof of Theorem V.

We conclude by observing that the same proof gives the following result.

Theorem VI Let $K(z,\zeta)$ be holomorphic in $z,\zeta$ for $\zeta \notin V_0 + z$ ($V_0$ is a fixed wedge s.t. $z + V_0$ lies above $\Gamma$ for all $z$) and such that

$$|K(z,\zeta)| \le \frac{C}{|z-\zeta|} \qquad \text{for } \zeta \notin V_1 + z$$

($V_1 \supset \bar{V}_0$ is a larger wedge).

If $\|\varphi'\|_\infty \le \delta_0$ then

$$\int_\Gamma k(z,\zeta) f(\zeta) d\zeta$$

is bounded on $L^2(\Gamma)$.

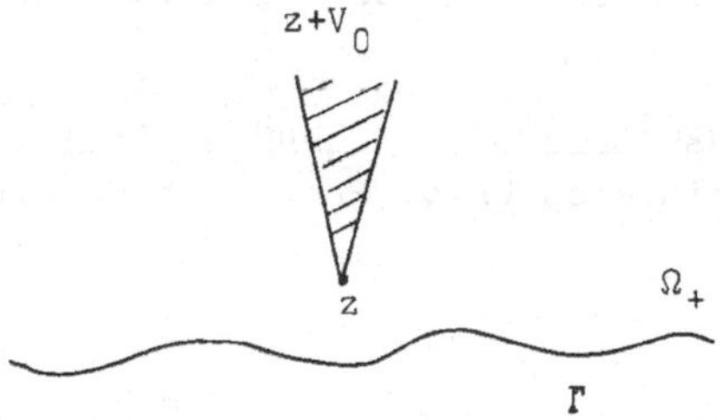


A natural example is $k_+(z) = \frac{1}{z^{1+i\gamma}}$ where $-\frac{\pi}{2} < \arg z < \frac{3\pi}{2}$.

A combination of this with the kernel corresponding to the determination $-\frac{3\pi}{2} < \arg z < \frac{\pi}{2}$ gives the boundedness on $L^2$ of the operator,

$$M_\gamma(f) = \int_\Gamma \frac{f(\zeta)}{|z-\zeta|^{1+i\gamma}} d\gamma.$$

It is a simple exercise using the Calderón Zygmund theory to show that these operators are also bounded on $L^p$, $1 < p < \infty$.

We conclude by observing that the operators studied so far in $\mathbb{R}^1$ admit natural extensions to $\mathbb{R}^n$. The so called rotation method permits one to deduce the corresponding estimate on $L^p(\mathbb{R}^n)$ and apply them to various questions in P.D.E, see [2], [3], [8].

REFERENCES


[1]. Calderón, A. P.,"Commutators of singular integral operators", Proc. Nat. Acad. Sci. U.S.A. 53 (1965) 1092-1099.

[2]. ______________, "Algebras of singular integral operators", Proc. Symp. Pure Math. 10 (1966) 18-55.

[3]. ______________,"Cauchy integrals on Lipschitz curves and related operators",Proc. Nat. Acad. Sci. U.S.A. 74 (1977) 1324-1327.

[4]. Coifman, R. and Meyer Y.,"On commutators of singular integrals and bilinear singular integrals",Trans. A.M.S. 212 (1975) 315-331.

[5]. ______________________, "Au dela des operateurs pseudo-differentiels", Asterisque 57 (1978).

[6]. Fefferman, C. and Stein, E.M.,"$H^p$ spaces of several variables", Acta. Math. 129 (1972) 137-193.

[7]. Kenig, C.E.,"$H^p$ spaces of Lipschitz domains",Thesis, U. of Chicago 1977.

[8]. Stein, E.M., Singular integrals and differentiability properties of functions, Princeton Univ. Press, Princeton, N.J. (1970).

# THE COMPLEX METHOD FOR INTERPOLATION OF OPERATORS ACTING ON FAMILIES OF BANACH SPACES

by

R.R. Coifman, R. Rochberg, and G. Weiss[1]
Washington University, St. Louis

M. Cwikel
Israel Institute of Technology

Y. Sagher
University of Illinois
at Chicago Circle

## §1. Introduction

It has not been sufficiently emphasized that many operators studied in analysis are particular values of analytic operator-valued functions, $z \to T_z$, defined on a domain $D \subset \mathbb{C}$. Moreover, it is often the case that the "natural" domain of $T_z$ is a Banach space $B_z$ that varies with $z$ in a smooth way. This is true, for example, for the Fourier transform. To see this consider the domain $D = \{z \in \mathbb{C}: |z| \le 1$ and $\operatorname{Re} z \ge 0\}$. For $z \in D$ let

$$\frac{1}{p(z)} = \operatorname{Re}\{\frac{1}{1+z}\} \quad \text{and} \quad \frac{1}{q(z)} = \operatorname{Re}\{\frac{z}{1+z}\} ;$$

we then define $T_z$ by describing its action on the Hermite polynomials $\{H_n\}$:

$$T_z: H_n \to z^n H_n .$$

It turns out that $T_z$ maps $L^{p(z)}(d\mu)$ into $L^{q(z)}(d\mu)$ with operator norm $\mathfrak{M}(z) = 1$ for $z \in D$, where $\mu$ is the measure on $\mathbb{R}$ defined by

$$d\mu(x) = \frac{1}{\sqrt{2\pi}} e^{-x^2/2} dx.$$

The functions $H_n(x)e^{-x^2/2}$, $n = 0,1,2,\ldots,$ form a complete orthogonal system of proper vectors of the Fourier transform, $F$, acting on $L^2(\mathbb{R},dx)$. The corresponding proper values are $i^n$. It follows, therefore, that, by appropriate multiplications by the function $e^{-x^2/2}$, we can identify $F$ and $T_i$. In fact, Beckner [1] showed by a change of variable argument, that when $z = iy$, $0 \le y \le 1$, the fact that $\mathfrak{M}(iy) = 1$ is equivalent to his remarkable sharp inequality

(1) This research was supported in part by NSF grant MCS 75-02411-A03 and NSF grant MCS 76-05789-A01.

(1.1) $$\|Ff\|_q \le A_p \|f\|_p,$$

where $p = p(iy)$, $q = q(iy)$ and $A_p^{\ell} = p^{1/p}/q^{1/q}$.

There are many other examples of such analytic families of operators. The fractional integrals and the irreducible representations of semi-simple Lie groups are well known analytic families of operators having features similar to the ones described above.

In view of the analytic nature of such families of operators, and the corresponding families of linear spaces they map from and into, it is reasonable to expect that each such family is completely determined by its values on the boundary of $D$. We shall see that this is, indeed, the case. Another way of looking at this property is in terms of the theory of interpolation of operators.

The current theory of interpolation involves the intermediate spaces between <u>two</u> given Banach spaces (the "end-point" or "boundary" spaces). The origin of this theory is the celebrated theorem of Riesz-Thorin involving operators acting on $L^p$-spaces on a measure space $(X,\mu)$ with values that are measurable functions on a measure space $(Y,\nu)$:

*Suppose $T$ is a linear operator mapping $L^{p_j}(X,\mu)$ into $L^{q_j}(Y,\nu)$ having operator norm $M_j < \infty$, $j = 0,1$. Then if $0 \le t \le 1$,*

$$\frac{1}{p} = \frac{1-t}{p_0} + \frac{t}{p_1} \quad \text{and} \quad \frac{1}{q} = \frac{1-t}{q_0} + \frac{t}{q_1},$$

*$T$ is well defined on $L^p(X,\mu)$ and maps the latter into $L^q(Y,\nu)$ with operator norm*

(1.2) $$M_t \le M_0^{1-t} M_1^t.$$

This result has been extended in many directions. E.M. Stein [6] considered analytic families of linear operators $\{T_z\}$ defined on the strip $\{z \in \mathbb{C} : 0 \le \text{Re } z \le 1\}$. He showed that appropriate boundedness assumptions of $T_{j+iy}$ mapping $L^{p_j}(X,\mu)$ into $L^{q_j}(Y,\nu)$, $j = 0,1$, implied the boundedness of $T_t$ as an operator from $L^p(X,\mu)$ into $L^q(Y,\nu)$ ($p$ and $q$ related to $t$ as above). Moreover, he obtained an inequality corresponding to (1.2). Shortly after Stein's paper was published there was an "explosion" of extensions of the Riesz-Thorin theorem. The principal goal of these extensions was to include more general Banach spaces (not just $L^p$ spaces) as the domains and ranges of the operator $T$. More explicitly, the problem posed by several investigators was the following one: Suppose $T$ is defined on two

Banach spaces $B_0$ and $B_1$ and it maps them boundedly into two Banach spaces $C_0$ and $C_1$, can one then construct "intermediate spaces" $B_t = [B_0,B_1]_t$ and $C_t = [C_0,C_1]_t$, for $0 \le t \le 1$, in such a way that $T$ maps $B_t$ into $C_t$ ?

As we mentioned above, all these results dealt with two "endpoint" or "boundary" spaces. It is our intention here to consider a situation (like the one we described in the beginning) where we are dealing with a continuum of Banach spaces, $B_\zeta$, associated with the points $\zeta$ of the boundary $\partial D$ of a domain $D \subset \mathbb{C}^n$. It is our purpose to construct "intermediate spaces" $B_z$, for each $z \in D$, in such a way that we obtain an "interpolation theory" for analytic families of linear operators, $\{T_z\}$, defined on $D$. The construction will be based on an extension of the "complex method" of A.P. Calderón [2].

This presentation consists of a considerably more detailed description of the results obtained in [3]. We wish to thank our collegues M. Taibleson and E. Wilson who made several very helpful suggestions concerning this paper.

## §2. Interpolation of operators acting on $L^p$-spaces

Before beginning the general construction of intermediate spaces we shall consider a generalization of the Riesz-Thorin theorem (and, also, of the theorem of E.M. Stein) which involves only $L^p$-spaces.

Let $D \subset \mathbb{C}$ be a domain and suppose that for each $z \in D$ there exists a linear operator $T_z$ mapping simple functions on a measure space $(X,\mu)$ into measurable functions on a measure space $(Y,\nu)$. We assume that $z \to \int_Y (T_z f)g \, d\nu$ is an analytic function on $D$, for each $f$ and $g$ simple. Suppose, further, that $\frac{1}{p(z)}$ and $\frac{1}{q(z)}$ are harmonic functions on $D$ with values in $[0,1]$. We let $\mathfrak{M}(z)$ denote the norm of $T_z$ as an operator from $L^{p(z)}(X,\mu)$ into $L^{q(z)}(Y,\nu)$. The interpolation result in question will follow from the following theorem:

Theorem (2.1). If $\mathfrak{M}(z) < \infty$ for $z \in D$ then $\log \mathfrak{M}(z)$ is subharmonic.

Proof: Fix $z_0 \in D$ and suppose $\overline{S_\rho(z_0)} = \{z \in \mathbb{C} : |z-z_0| \le \rho\} \subset D$. Let $a(z)$ and $b(z)$ be the unique analytic functions in $\overline{S_\rho(z_0)}$ whose real parts are $\frac{1}{p(z)}$ and $1 - (\frac{1}{q(z)}) \equiv \frac{1}{r(z)}$ (respectively) and such that $a(z_0)$ and $b(z_0)$ are real numbers. Choose simple functions $f$ and $g$ such that

$$(2.2)\qquad \int_X |f|^{p(z_0)}\,d\mu = 1 = \int_Y |g|^{r(z_0)}\,d\nu .$$

Let $f' = f/|f|$ when $f$ is not zero and, otherwise, put $f' = 0$; define $g'$ similarly. We then put

$$f_z = |f|^{a(z)p(z_0)} f' \quad\text{and}\quad g_z = |g|^{b(z)r(z_0)} g' .$$

An immediate calculation, which makes use of (2.2), yields

$$(2.3)\qquad \|f_z\|_{p(z)} = 1 = \|g_z\|_{r(z)} .$$

Moreover, $f_{z_0} = f$ and $g_{z_0} = g$. The function

$$F(z) = \int_Y (T_z f_z) g_z \, d\nu$$

is analytic in $\overline{S_\rho(z_0)}$ (it is the sum of terms of the form $e^{\varphi(z)} \int_Y (T_z \chi_E)\chi_F \, d\nu$, where $\varphi$ is an analytic function and $\chi_E$, $\chi_F$ are characteristic functions of measurable subsets of $X$ and $Y$, of finite measure). Thus, $\log|F(z)|$ is subharmonic. Consequently, making use of this fact and (2.3) we have:

$$\log\left|\int_Y (T_{z_0} f) g \, d\nu\right| = \log|F(z_0)| \le \frac{1}{2\pi}\int_0^{2\pi} \log|F(z_0+\rho e^{i\theta})|\,d\theta$$
$$\le \frac{1}{2\pi}\int_0^{2\pi} \log \mathfrak{M}(z_0+\rho e^{i\theta})\,d\theta .$$

Now, taking the supremum over all simple $f$ and $g$ satisfying $\|f\|_{p(z_0)} = 1 = \|g\|_{r(z_0)}$, we obtain

$$(2.4)\qquad \log \mathfrak{M}(z_0) \le \frac{1}{2\pi}\int_0^{2\pi} \log \mathfrak{M}(z_0+\rho e^{i\theta})\,d\theta .$$

This establishes the subharmonicity of $\log \mathfrak{M}(z)$ and Theorem (2.1).

Let us now make several remarks concerning this theorem. We first note that the Riesz-Thorin theorem is a very special case of this result. Suppose $T$ satisfies the hypotheses preceding (1.2). Let $D = \{z \in \mathbb{C} : |z| < 1\}$ be the unit disc and $\frac{1}{p(z)}$, $\frac{1}{q(z)}$ the solution of the Dirichlet problem with boundary data

$$\frac{1}{p(e^{i\theta})} = \begin{cases} \dfrac{1}{p_0} & \text{if } -\pi \le \theta \le \pi(1-2t) \\[2mm] \dfrac{1}{p_1} & \text{if } \pi(1-2t) < \theta < \pi \end{cases}$$

and

$$\frac{1}{q(e^{i\theta})} = \begin{cases} \dfrac{1}{q_0} & \text{if} \quad -\pi \le \theta \le \pi(1-2t) \\[2ex] \dfrac{1}{q_1} & \text{if} \quad \pi(1-2t) < \theta < \pi \end{cases}$$

respectively. Then the argument we gave for inequality (2.4) can clearly be extended, in this case, to the situation where $z_0 = 0$ and $\rho = 1$; thus, letting $T_z \equiv T$

$$\log \mathfrak{N}(0) \le \frac{1}{2\pi} \int_0^{2\pi} \log \mathfrak{N}(e^{i\theta}) \, d\theta = (1-t) \log M_0 + t \log M_1.$$

since, $\frac{1}{p(0)} = \frac{1-t}{p_0} + \frac{t}{p_1}$ and $\frac{1}{q(0)} = \frac{1-t}{q_0} + \frac{t}{q_1}$, we obtain (1.2) from the last inequality by taking exponentials.

Similarly, Stein's theorem on interpolation of analytic families of operators can be obtained immediately from theorem (2.1). In order to see this we can take the same domain $\bar{D} = \{z = x+iy \in \mathbb{C} : 0 \le x \le 1\}$ he considers in [6]. His condition of admissible growth is precisely the one needed to construct the least harmonic majorant of the subharmonic function $\log \mathfrak{N}(z)$, which is used to obtain the bounds he has for the norm of the operator $T_t$, $0 \le t \le 1$ (see (2.4) in [6]).

The family of operators $T_z : H_n \to z^n H_n$, described in the beginning of §1, offers good examples for the application of theorem (2.1). Let $D$ be the semi-disc bounded by the vertical line segment $\{z = iy : -1 \le y \le 1\}$ and the semicircle $\{z = e^{i\theta} : -\frac{\pi}{2} \le \theta \le \frac{\pi}{2}\}$. If we assume Beckner's result that $T_{iy} : L^{1+y^2}(d\mu) \to L^{(1+y^2)/y^2}(d\mu)$ is a bounded operator of norm 1 and the trivial fact that $T_{e^{i\theta}} : L^2(d\mu) \to L^2(d\mu)$ is an isometry, then theorem (2.1) gives us the result $\mathfrak{N}(z) \le 1$. When $z = x$ is real and in $D$, the fact that $\mathfrak{N}(x) = 1$ was shown by L. Gross. The subharmonicity of $\mathfrak{N}(z)$, then, gives us the fact that $T_z$ has operator norm 1. We could, also, use the Beckner result to obtain information about the operators $T_z$ for $\{z = x+iy : |z| \le 1, -1 \le x \le 0\}$. We merely consider the reflection $\tilde{D}$ of $D$ in the y-axis, let $\frac{1}{p(z)} = \text{Re}\{\frac{1}{1-z}\}$, $\frac{1}{q(z)} = \text{Re}\{\frac{z}{z-1}\}$ and, again, use Beckner's result on the y-axis segment of $\partial \tilde{D}$ and the isometry property of $T_{e^{i\theta}} : L^2(d\mu) \to L^2(d\mu)$, $\frac{\pi}{2} \le \theta \le \frac{3\pi}{2}$, in order to obtain the boundedness of the operators

$$T_z : L^{p(z)}(d\mu) \to L^{q(z)}(d\mu)$$

for $z \in \tilde{D}$. In particular, we obtain the fact that $T_x\colon L^{1+|x|}(d\mu) \to L^{(1+|x|)/|x|}(d\mu)$ is an operator of norm not exceeding 1 for $-1 \le x \le 1$. It is tempting to try to use this last result in order to interpolate in the upper semi-disc and see if Beckner's result is obtained from (2.1). This appears not to be the case(2).

## §3. The construction of intermediate spaces

We have just seen how "intermediate spaces" can be constructed in a domain $D$ when an appropriate family of $L^p$ spaces is assigned to the points of the boundary $\partial D$. More precisely, if we are given a measurable function $\frac{1}{p(\zeta)}$, $\zeta \in \partial D$, $0 \le \frac{1}{p(\zeta)} \le 1$, and a solution, $\frac{1}{p(z)}$, of the Dirichlet problem with this boundary data, then the spaces $L^{p(z)}$ are the intermediate spaces for which interpolation results of the type (2.1) can be obtained. We would now like to extend this method to the case where the boundary spaces are general Banach spaces. The first difficulty that arises when one attempts to attack this problem is that there is no real hope to have well defined, non-trivial intermediate spaces if the boundary Banach spaces have a null, or "small", intersection. The fact that the simple functions form a substantial subspace of the $L^p$ spaces allows us, for example, to consider a family of operators $\{T_z\}$ having a common domain. This problem arises even in the theory of intermediate spaces $[B_0,B_1]_t$ between two Banach spaces $B_0$ and $B_1$. We shall avoid all these difficulties by considering all our Banach spaces to be $\mathbb{C}^n$ endowed with general norms. We shall see that the interpolation estimates we obtain do not depend on the dimension $n$. For most applications this is all we need, since a large class of Banach spaces has the property that they are appropriate limits of finite dimensional subspaces. Also, we shall consider, for the most part, the domain $D$ to be the unit disc. The extension of our method to more general domains is routine.

We assume, then, that for each $\theta \in [0,2\pi)$ we have a norm $|\ |_{e^{i\theta}}$ on $\mathbb{C}^n$ such that $\theta \to |v|_{e^{i\theta}}$ is a measurable function for each $v \in \mathbb{C}^n$ and that

$$(3.1) \qquad k_1(\theta)\|v\| \le |v|_{e^{i\theta}} \le k_2(\theta)\|v\|,$$

---

(2) See [3] for further discussion concerning (2.1) and Beckner's result. The inequalities obtained by Weissler [7] could also be similarly studied in connection with (2.1).

where $\|v\| = (\sum_{k=1}^{n} |v_k|^2)^{1/2}$ is the usual Euclidean norm and $\log k_j(\theta)$, $j = 1,2$, is integrable on $[0,2\pi)$. We have, therefore, a family of Banach spaces $B_{e^{i\theta}} = (\mathbb{C}^n, |\ |_{e^{i\theta}})$ assigned to the points of the boundary $\partial D$. Our task will be to construct an appropriate family $B_z = (\mathbb{C}^n, |\ |_z)$ for $z \in D$.

For $z = re^{i\varphi} \in D$ let

$$h_z(\theta) = \frac{1}{2\pi}\frac{1+ze^{-i\theta}}{1-ze^{-i\theta}} = \frac{1}{2\pi}\frac{1-r^2}{1-2r\cos(\varphi-\theta)+r^2} + i\frac{1}{\pi}\frac{r\sin(\varphi-\theta)}{1-2r\cos(\varphi-\theta)+r^2}$$

$$= P_z(\theta) + Q_z(\theta).$$

$h_z(\theta)$ is usually called the Herglotz kernel, $P_z(\theta)$ and $Q_z(\theta)$ are the Poisson and conjugate Poisson kernels associated with the unit disc. We let

$$W_j(z) = \exp\{\int_0^{2\pi} h_z(\theta)\log k_j(\theta)\,d\theta\}$$

for $j = 1,2$ and $z \in D$. The integrability of $\log k_j(\theta)$ assures us that the functions $W_j(z)$ are well defined, never vanishing analytic functions on $D$ (in fact, $\log W_j(z) \in H^p(D)$ for $0 < p < 1$). We let the symbol $z \rhd e^{i\theta}$ denote a general non-tangential approach of $z \in D$ to the boundary point $e^{i\theta}$. From classical $H^p$-space theory (see [10], Ch. VII) we have

$$\text{(3.2)} \qquad \lim_{z \rhd e^{i\theta}} |W_j(z)| = k_j(\theta),$$

almost everywhere. For $p > 0$ Let $H_j^p = H_j^p(D;\mathbb{C}^n)$, $j = 1,2$, consist of all $\mathbb{C}^n$-valued analytic functions on $D$ satisfying

$$\|F\|_p^j \equiv \sup_{0\le r<1} \left(\int_0^{2\pi} \|W_j(re^{i\theta})F(re^{i\theta})\|^p \frac{d\theta}{2\pi}\right)^{1/p} < \infty.$$

Thus, $F = (f_1,f_2,\ldots,f_n) \in H_j^p$ if and only if $W_j f_k \in H^p(D)$ for $k = 1,2,\ldots,n$. It follows that the non-tangential limits $F(e^{i\theta})$ exist a.e. (by (3.2) we have $\lim_{z \rhd e^{i\theta}} |W_j(z)|^{-1} = \frac{1}{k_j(\theta)} < \infty$ a.e.). We can, therefore, introduce the spaces

$$H_\#^p = \{F \in H_1^p : \ |||F|||_p = \left(\int_0^{2\pi} |F(e^{i\theta})|_{e^{i\theta}}^p \frac{d\theta}{2\pi}\right)^{1/p} < \infty\}.$$

Thus, $H_\#^p \subset H_1^p$ and, because of (3.1) and (3.2),

$$\|F\|_p^1 \le |||F|||_p .$$

It will be useful to introduce the following norms on $H^p_\#$: for each $z_0 \in D$ we let

$$|||F|||_{p,z_0} = \left( \int_0^{2\pi} |F(e^{i\theta})|^p_{e^{i\theta}} P_{z_0}(\theta)\, d\theta \right)^{1/p}.$$

We then have $|||F|||_p = |||F|||_{p,0}$ and this norm is equivalent to $|||F|||_{p,z_0}$. Moreover, $|||F|||_{\infty,z_0}$ is independent of $z_0 \in D$.

If $g \in H^p(D)$ it is easy to check that $|g(z_0)| \le \dfrac{c\|g\|_p}{(1-|z_0|)^{1/p}}$, for $z_0 \in D$ and $c$ an appropriate constant independent of $g$. Applying this inequality to each component of $W_jF$, for $F \in H^p_j$, we obtain

$$(3.3) \qquad \|F(z_0)\| \le c\|F\|^j_p / (1-|z_0|)^{1/p}|W_j(z_0)|$$

for each $z_0 \in D$. From this we obtain:

Lemma (3.4). If $K$ is a compact subset of $D$ and $F \in H^p_j$, then there exists a constant $c_K = c^j_K$ such that

$$\|F(z)\| \le c_K\|F\|^j_p$$

for all $z \in K$.

We can now introduce a "norm" on $\mathbb{C}^n$ associated with each $z_0 \in D$. Its definition gives the impression that it depends on the index $p > 0$. We shall see that this is not the case and that it does satisfy the properties we are seeking for the intermediate space associated with $z_0$. For $v \in \mathbb{C}^n$ we let

$$|v|_{z_0,p} = \inf\{|||F|||_{p,z_0} : F \in H^p_\#, F(z_0) = v\}.$$

Lemma (3.5). $|v|_{z_0,p} = 0$ if and only if $v = 0$.

Proof: Clearly $v = 0$ implies $|v|_{z_0,p} = 0$. If, on the other hand, $|v|_{z_0,p} = 0$ we can find, given $\varepsilon > 0$, an $F \in H^p_\#$ such that $F(z_0) = v$ and $|||F|||_{p,z_0} < \varepsilon$. By (3.4), with $K = \{z_0\}$, we then have

$$\|v\| = \|F(z_0)\| \le c_{\{z_0\}}\|F\|^1_p \le c_{\{z_0\}} |||F|||_p \le c' |||F|||_{p,z_0} < c'\varepsilon.$$

Since $\varepsilon$ can be arbitrarily small, $\|v\| = 0$ and, thus, $v = 0$.

If $1 \le p$ it follows immediately from (3.5), and its definition, that $|\ |_{z_0,p}$ is a norm. Since, as we stated above, $|\ |_{z_0,p}$ is independent of $p$, we obtain a norm on $\mathbb{C}^n$ also when $0 < p < 1$. Before showing this independence result, however, we prove the following:

Lemma (3.6). $H^p_\#$ is complete.

Proof: Suppose $\{F_k\}$ is a Cauchy sequence in $H^p_\#$. Then, since $\|F_k - F_m\|^1_p \le |||F_k - F_m|||_p$, it follows that $\{W_1 F_k\}$ is a Cauchy sequence in $H^p(D;\mathbb{C}^n)$ (the space of $\mathbb{C}^n$-valued functions having each component in classical $H^p(D)$). From the completeness of the latter space we are assured of the existence of an $F \in H^p_1$ such that $\lim_{k\to\infty} \|F - F_k\|^1_p = 0$. Relabelling, if necessary, we can assume that $\lim_{k\to\infty} F_k(e^{i\theta}) = F(e^{i\theta})$ a.e. Thus, using Fatou's lemma, we have:

$$
\begin{aligned}
|||F - F_k|||^p_p &= \int_0^{2\pi} |F(e^{i\theta}) - F_k(e^{i\theta})|^p_{e^{i\theta}} \frac{d\theta}{2\pi} \\
&= \int_0^{2\pi} \lim_{m\to\infty} |F_m(e^{i\theta}) - F_k(e^{i\theta})|^p_{e^{i\theta}} \frac{d\theta}{2\pi} \\
&\le \underline{\lim}_{m\to\infty} \int_0^{2\pi} |F_m(e^{i\theta}) - F_k(e^{i\theta})|^p_{e^{i\theta}} \frac{d\theta}{2\pi} \\
&= \underline{\lim}_{m\to\infty} |||F_m - F_k|||^p_p .
\end{aligned}
$$

But the last expression is as small as we wish if $k$ is large enough. Thus, $F = \lim_{k\to\infty} F_k$ in $H^p_\#$, showing the desired completeness.

We shall now state and prove the following basic result concerning the norms $|\ |_{z_0,p}$ we introduced:

Theorem I. If $v \in \mathbb{C}^n$, $z_0 \in D$ and $0 < p \le \infty$, then

$$|v|_{z_0,p} = \inf\{|||G|||_\infty : G \in H^p_\#,\ G(z_0) = v,\ |G(e^{i\theta})|_{e^{i\theta}} = \text{const. a.e.}\}$$

is independent of $p$. Furthermore, there exists an extremal function $F = F_{z_0,v} \in \{G \in H^\infty_\# : G(z_0) = v,\ |G(e^{i\theta})|_{e^{i\theta}} = \text{const. a.e.}\}$ such that $|v|_{z_0} = |v|_{z_0,p} = |F(e^{i\theta})|_{e^{i\theta}}$ a.e.

Proof: Let $E = E^v_{p,z_0} = \{F \in H^p_\# : F(z_0) = v\}$. We first show $E \ne \emptyset$ (observe that our hypotheses do not imply that the constant function

$F(z_0) \equiv v$ belongs to $H^p_\#$). Let $F(z) = \frac{W_2(z_0)}{W_2(z)} v$. Clearly, $F(z_0) = v$. To check that $F \in H^p_\#$ we note that

$$|W_1(z)/W_2(z)| = \exp\left(\int_0^{2\pi} P_z(\theta) \log[k_1(\theta)/k_2(\theta)]\, d\theta\right) \leq 1$$

(because $\log[k_1(\theta)/k_2(\theta)] \leq 0$). It follows that $W_1(z)F(z) = \frac{W_1(z)}{W_2(z)} W_2(z_0)\, v$ belongs to $H^\infty(D;\mathbb{C}^n) \subset H^p(D;\mathbb{C}^n)$. Thus, $F \in H^p_1$. Moreover, $F(e^{i\theta}) = \frac{W_2(z_0)}{W_2(e^{i\theta})} v$ a.e.; consequently,

$$|||F|||_{p,z_0} = \left(\int_0^{2\pi} P_{z_0}(\theta) |W_2(z_0)|^p [|v|_{e^{i\theta}}/k_2(\theta)]^p\, d\theta\right)^{1/p}$$

$$\leq |W_2(z_0)|\, \|v\| < \infty .$$

This shows that $F \in E$.

We now choose a sequence $\{F_k\} \subset E$ such that $\lim_{k\to\infty} |||F_k|||_{p,z_0} = |v|_{z_0,p}$ and let

$$\Phi_k(z) = \exp\left\{\int_0^{2\pi} [h_{z_0}(\theta) - h_z(\theta)] \log|F_k(e^{i\theta})|_{e^{i\theta}}\, d\theta\right\}.$$

Put $K_k = \Phi_k F_k$. We also introduce the notation

$$|||G|||_{0,z_0} = \exp\left\{\int_0^{2\pi} \log|G(e^{i\theta})|_{e^{i\theta}} P_{z_0}(e^{i\theta})\, d\theta\right\}.$$

$|||G|||_{0,z_0}$ is the <u>geometric mean</u> of $G$ with respect to the probability measure $P_{z_0}(e^{i\theta})\, d\theta$ on $[0,2\pi)$. If, for some $p > 0$, $|||G|||_{p,z_0} < \infty$ then

$$\text{(3.7)} \qquad \lim_{p\to 0} |||G|||_{p,z_0} = |||G|||_{0,z_0}$$

(see Chapter 6 of [ 5 ]). We then have,

$$\|W_1(z)K_k(z)\|$$

$$= |W_1(z)|\,|\Phi_k(z)|\,\|F_k(z)\|$$

$$= |||F_k|||_{0,z_0} \exp \int_0^{2\pi} P_z(\theta) \log \frac{k_1(\theta)}{|F_k(e^{i\theta})|_{e^{i\theta}}}\, d\theta\}\ \|F_k(z)\|$$

$$\leq \; |||F_k|||_{0,z_0} \, \|F_k(z)\| \, \exp \int_0^{2\pi} P_z(\theta) \log \frac{1}{\|F_k(e^{i\theta})\|} \, d\theta$$

$$= \; |||F_k|||_{0,z_0} \, \|W_1(z)F_k(z)\| \; / \; \exp\{ \int_0^{2\pi} P_z(\theta) \log\|k_1(\theta)F_k(e^{i\theta})\| \, d\theta \}.$$

It follows easily from classical $H^p$-theory that the last term does not exceed $|||F_k|||_{0,z_0}$. This shows that $\|W_1(z)K_k(z)\| \leq |||F_k|||_{0,z_0} \leq |||F_k|||_{p,z_0}$. Thus, $K_k \in H_1^\infty(D;\mathbb{C}^n)$. Moreover, $|K_k(e^{i\theta})|_{e^{i\theta}} = |||F_k|||_{0,z_0}$ a.e. This shows that $K_k \in H_\#^\infty \subset H_\#^p$. Since $K_k(z_0) = v$ we must have

$$|v|_{z_0,p} \;\leq\; |||K_k|||_{p,z_0} \;=\; |||F_k|||_{0,z_0} \;\leq\; |||F_k|||_{p,z_0} .$$

But $|||F_k|||_{p,z_0} \to |v|_{z_0,p}$ as $k \to \infty$. It follows that

$$|v|_{z_0,p} \;=\; \inf\{ |||G|||_\infty : \; G \in H_\#^\infty, \; G(z_0) = v, \; |G(e^{i\theta})|_{e^{i\theta}} = \text{const. a.e.}\}$$

and the first part of theorem I is established.

In order to show the last part let us accept the fact (which will be shown in the proof of theorem II) that, for $1 < p < \infty$, $H_\#^p$ is a closed subspace of a reflexive Banach space. Since the set $\mathcal{E}$ is a convex and closed[2] subset of this reflexive Banach space it contains an element $F$ of minimal norm (see pg. 244 of [9]). Thus, $|v|_{z_0} = |||F|||_{p,z_0}$. As before, let

$$\Phi(z) \;=\; \exp\{ \int_0^{2\pi} [h_{z_0}(\theta) - h_z(\theta)] \log|F(e^{i\theta})|_{e^{i\theta}} \, d\theta \}$$

and put $K = \Phi F$. The argument we just gave (with $F_k = F$ for all $k$) implies

$$|v|_{z_0} \;\leq\; |||K|||_{p,z_0} \;=\; |||F|||_{0,z_0} \;\leq\; |||F|||_{p,z_0} \;=\; |v|_{z_0}.$$

Thus, we must have equalities in all the last inequalities. But, on a probability space $(M,\mu)$, if $\exp \int_M \log|f| \, d\mu = (\int_M |f|^p \, d\mu)^{1/p}$ for some $p > 0$, we must have $|f(x)| = \text{const. a.e.}$ (see Ch. 6 of [ 5 ]). This shows that $|F(e^{i\theta})|_{e^{i\theta}} = |v|_{z_0}$ a.e. and theorem I is proved.

Let us make some observations concerning this proof. The argument we just gave showed:

---

(2) Lemmas (3.4) and (3.6) can be used to show $\mathcal{E}$ is closed.

Corollary (3.8). If $F$ is an extremal function in $E = E^{v}_{p,z_0}$ then $F \in H^{\infty}_{\#}$ and $|F(e^{i\theta})|_{e^{i\theta}} = |v|_{z_0}$ a.e.

We also showed, in the argument that proved that $E \neq \emptyset$, that $F(z) = \frac{W_2(z_0)}{W_2(z)} v \in E$. But this implies that

$$|v|_{z_0} \leq |||F|||_{\infty} = \underset{\theta\in[0,2\pi)}{\text{ess. sup.}} \left\{ \frac{|W_2(z_0)|}{k_2(\theta)} |v|_{e^{i\theta}} \right\} \leq |W_2(z_0)| \, \|v\|.$$

That is,

$$|v|_{z_0} \leq |W_2(z_0)| \, \|v\|. \tag{3.9}$$

Our next task will be to investigate the question of how the dual spaces of the intermediate spaces we have constructed are related to the dual spaces of the original boundary spaces. In order to do this we introduce some notation and make some elementary observations. If $N$ is a norm on $\mathbb{C}^n$, we define $N^*(w) = \sup\{|\langle v,w\rangle|: N(v) = 1\}$ for each $w \in \mathbb{C}^n$, where $\langle v,w\rangle = \sum_{j=1}^{n} v_j w_j$. $N^*$ is, then, also a norm, the <u>dual norm</u> (associated with $N$). Thus the linear functional $L_w: v \to \langle v,w\rangle$ is well defined on $(\mathbb{C}^n,N)$ and has norm $\leq N^*(w)$. By a simple compactness argument one can show that the norm of $L_w$ is actually equal to $N^*(w)$. By a dimension argument one can also show that every linear functional on $(\mathbb{C}^n,N)$ is of this form. This justifies the identification of the dual, $(\mathbb{C}^n,N)^*$ of $(\mathbb{C}^n,N)$ with $(\mathbb{C}^n,N^*)$.

<u>Lemma (3.10).</u> $N^{**}(v) = N(v)$ for all $v \in \mathbb{C}^n$.

<u>Proof</u>: Clearly, $N^{**}(v) \leq N(v)$. To show the opposite inequality we construct, using the Hahn-Banach theorem, a linear functional $L$ of norm 1 such that $L(v) = N(v)$. Since there exists $w \in \mathbb{C}^n$, with $N^*(w) = 1$, such that $L = L_w$ we have $N(v) = L(v) = L_w(v) = \langle v,w\rangle \leq N^{**}(v)N^*(w) = N^{**}(v)$.

The following lemma follows easily from (3.10) and the definitions:

<u>Lemma (3.11).</u> $k\|v\| \leq N(v) \leq K\|v\|$ if and only if $K^{-1}\|v\| \leq N^*(v) \leq k^{-1}\|v\|$ (for all $v \in \mathbb{C}^n$, $k$, $K$ positive constants).

Now suppose we have the situation we encountered in theorem I: a family of norms $\{|\ |_{e^{i\theta}}\}$ is associated with the points of the boundary of $D$ which satisfies (3.1). Then, by (3.11), the dual norms satisfy

$$(3.12) \qquad \frac{1}{k_2(\theta)} \|v\| \leq |v|^*_{e^{i\theta}} \leq \frac{1}{k_1(\theta)} \|v\|$$

and the functions $\log \frac{1}{k_j(\theta)}$ are integrable, by hypothesis. That is, we can apply the method of theorem I to the boundary space $(\mathbb{C}^n, |\ |^*_{e^{i\theta}})$ and obtain corresponding intermediate spaces.

More generally, let us introduce the following notation. Suppose $D$ is a domain in $\mathbb{C}$ and $\{B_\zeta\} = \{(\mathbb{C}^n, |\ |_\zeta\}$, $\zeta \in \partial D$, is a family of Banach spaces for which the construction of theorem I can be performed. We then denote by $[B_\zeta]_z$ the intermediate space obtained at $z \in D$.

<u>Theorem II</u> (The duality theorem). $[B^*_\zeta]_z = [B_\zeta]^*_z$, for all $z \in D$.

<u>Proof</u>: For simplicity we give the argument for $D$ the unit disc and $z = 0$; the modifications required for other $z \in D$ are obvious. Let

$$L^2_\# = \{f \text{ defined on } \partial D,\ \mathbb{C}^n\text{-valued and measurable:}$$
$$|||f|||_2 \equiv \left( \int_0^{2\pi} |f(e^{i\theta})|^2_{e^{i\theta}} \frac{d\theta}{2\pi} \right)^{1/2} < \infty\}.$$

(This is <u>not</u> a Hilbert space in general!) Also, let

$$L^2_2 = \{f \text{ defined on } \partial D,\ \mathbb{C}^n\text{-valued and measurable:}$$
$$\|f\|^{(2)}_2 = \left( \int_0^{2\pi} \|k_2(\theta) f(e^{i\theta})\|^2 \frac{d\theta}{2\pi} \right)^{1/2} < \infty\}.$$

This last space is a Hilbert space; hence, it is self-dual. It will be convenient for us, however, to identify the dual as the space

$$L^{2*}_2 = \{g \text{ defined on } \partial D,\ \mathbb{C}^n\text{-valued and measurable:}$$
$$\|g\|^{(2)*}_2 = \left( \int_0^{2\pi} \left\| \frac{1}{k_2(\theta)} g(e^{i\theta}) \right\|^2 \frac{d\theta}{2\pi} \right)^{1/2} < \infty\}$$

via the correspondence $g \leftrightarrow L_g$, where $L_g(f) = \int_0^{2\pi} \langle f,g \rangle \frac{d\theta}{2\pi}$. It is easy to check that the norm of the linear functional $L_g$, $\|L_g\|_{L^{2*}_2}$, equals $\|g\|^{(2)*}_2$.

Clearly, $|||f|||_2 \leq \|f\|^{(2)}_2$; thus, $L^2_2 \subset L^2_\#$. If $f \in L^2_\#$, let

$f_n = \psi_n f$ where

$$\psi_n(\theta) = \begin{cases} 1 & \text{if } k_2(\theta)\|f(e^{i\theta})\| \le n \\ \dfrac{n}{k_2(\theta)\|f(e^{i\theta})\|} & \text{otherwise .} \end{cases}$$

Thus, $0 \le \psi_n(\theta) \le \psi_{n+1}(\theta) \le 1$ and, consequently, $\|f_n(e^{i\theta})k_2(\theta)\| = \psi_n(\theta)\|k_2(\theta)f(e^{i\theta})\| \le n$. This shows $k_2 f_n \in L_2^2$ for each $n$. Hence, by the dominated convergence theorem,

$$|||f_n - f|||_2^2 = \int_0^{2\pi} [1-\psi_n(\theta)]^2 |f(e^{i\theta})|^2_{e^{i\theta}} \frac{d\theta}{2\pi}$$

tends to $0$ as $n \to \infty$. This shows that $L_2^2$ is dense in $L_{\#}^2$.

Now suppose $L$ is a bounded linear functional on $L_{\#}^2$ and $\|L\|$ denotes its norm. The restriction of $L$ to $L_2^2$ is then a bounded linear functional with norm $\le \|L\|$. Thus, there exists $g \in L_2^{2*}$ such that $L = L_g$ on $L_2^2$. We claim that

$$(3.13) \qquad |||g|||_2^* \equiv \left( \int_0^{2\pi} (|g(e^{i\theta})|^*_{e^{i\theta}})^2 \frac{d\theta}{2\pi} \right)^{1/2} \le \|L\|.$$

In order to see this we choose a measurable function $\tilde{f}$ which satisfies $|\tilde{f}(e^{i\theta})|_{e^{i\theta}} = 1$ and $|g(e^{i\theta})|^*_{e^{i\theta}} = \langle \tilde{f}(e^{i\theta}), g(e^{i\theta})\rangle$. Also let $\varphi$ be a complex valued function satisfying $\int_0^{2\pi} |\varphi(\theta)|^2 \frac{d\theta}{2\pi} = 1$ and put $f(e^{i\theta}) = \varphi(\theta)\tilde{f}(e^{i\theta})$. Then $f \in L_{\#}^2$ and $|||f|||_2 = 1$. Put $f_n = \psi_n f$, where $\psi_n$ is the function defined above. Thus

$$\begin{aligned} L(f_n) = L_g(f_n) &= \int_0^{2\pi} \psi_n(\theta)\langle f(e^{i\theta}), g(e^{i\theta})\rangle \frac{d\theta}{2\pi} \\ &= \int_0^{2\pi} \psi_n(\theta)\varphi(\theta)|g(e^{i\theta})|^*_{e^{i\theta}} \frac{d\theta}{2\pi} . \end{aligned}$$

But, by the monotone convergence theorem of Lebesgue, the last expression tends to $\int_0^{2\pi} \varphi(\theta)|g(e^{i\theta})|^*_{e^{i\theta}} \frac{d\theta}{2\pi} = \int_0^{2\pi} \langle f(e^{i\theta}), g(e^{i\theta})\rangle \frac{d\theta}{2\pi}$. On the other hand, since $\lim_{n\to\infty} |||f_n - f|||_2 = 0$ and $L$ is a continuous linear functional on $L_{\#}^2$, we must have $\lim_{n\to\infty} L(f_n) = L(f)$. Thus,

$$L(f) = \int_0^{2\pi} \varphi(\theta)|g(e^{i\theta})|^*_{e^{i\theta}} \frac{d\theta}{2\pi} = \int_0^{2\pi} <f(e^{i\theta}),g(e^{i\theta})> \frac{d\theta}{2\pi} .$$

Taking the supremum of all these expressions over all $\varphi \in L^2(0,2\pi)$ with $\|\varphi\|_2 = 1$ we obtain (writing $f = f_\varphi$)

$$|||g|||_2^* = \sup_\varphi L(f_\varphi) \le \sup_{|||h|||_2 \le 1} |L(h)| = \|L\|.$$

This establishes (3.13).

For any $f \in L^2_\#$

$$\int_0^{2\pi} |<f(e^{i\theta}),g(e^{i\theta})>| \frac{d\theta}{2\pi} \le \int_0^{2\pi} |f(e^{i\theta})|_{e^{i\theta}} |g(e^{i\theta})|^*_{e^{i\theta}} \frac{d\theta}{2\pi}$$

$$\le |||f|||_2 \; |||g|||_2^* < \infty.$$

Again writing $f_n = \psi_n f$ we have, using the dominated convergence theorem,

$$L(f) = \lim_{n\to\infty} L(f_n) = \lim_{n\to\infty} L_g(f_n) = \lim_{n\to\infty} \int_0^{2\pi} \psi_n(\theta)<f(e^{i\theta}),g(e^{i\theta})> \frac{d\theta}{2\pi}$$

$$= \int_0^{2\pi} <f(e^{i\theta}),g(e^{i\theta})> \frac{d\theta}{2\pi} .$$

Thus, $g$ represents $L$ on $L^2_\#$ and $|L(f)| \le |||g|||_2^* \; |||f|||_2$. It follows that $\|L\| \le |||g|||_2^*$. Consequently,

(3.14) $$\|L\| = |||g|||_2^* .$$

Now let us choose $v \in \mathbb{C}^n$ and let $L_v$ be the linear functional on $H^2_\#$ defined by $L_v(F) = <F(0),v>$. Then $|L_v(F)| \le |F(0)|_0 |v|_0^* \le |||F|||_2 \, |v|_0^*$ and it follows that $\|L_v\|_{(H^2_\#)^*} \le |v|_0^*$. By considering the boundary values of the functions in $H^2_\#$ we can consider the latter to be a closed subspace of $L^2_\#$.(3) By the Hahn-Banach theorem we can extend $L_v$ to a linear functional $L$ on $L^2_\#$ without increasing its

(3) The fact that $H^2_\#$ is closed follows from (3.6). The argument establishing (3.14) gives us an identification of the dual, $(L^2_\#)^*$, of $L^2_\#$ as the space of all $g$ such that $|||g|||_2^* < \infty$. Since $( |||f|||_2^* )^* = |||f|||_2$ it follows that $L^2_\#$ is reflexive. Recall that in the proof of theorem I we used the fact that $H^2_\#$ is a closed subspace of a reflexive Banach space (this argument extends to other indices $p$, $1 < p < \infty$).

norm. Let $g \in (L^2_\#)^*$ be the representative of L. Define $F_k(z) = (z^k/W_2(z))u$, for $k = 0,1,\ldots$ and $u \in \mathbb{C}^n$. Then,

$$L_v(F_k) = \langle F_k(0),v\rangle = \begin{cases} \langle u,v\rangle/W_2(0) & \text{if } k = 0 \\ 0 & \text{if } k > 0. \end{cases}$$

On the other hand, $L_v(F_k) = \frac{1}{2\pi}\int_0^{2\pi} \left\langle u, \frac{g(e^{i\theta})}{W_2(e^{i\theta})}\right\rangle e^{ik\theta}\, d\theta$. It follows that the Fourier series of the components of $g/W_2$ are of power series type. By (3.11)

$$\left\| \frac{1}{W_2(e^{i\theta})} g(e^{i\theta})\right\| = \frac{1}{k_2(\theta)} \|g(e^{i\theta})\| \leq |g(e^{i\theta})|^*_{e^{i\theta}}.$$

Consequently $\left\langle u, \frac{g(z)}{W_2(z)}\right\rangle$ is an analytic function in $H^2(D)$. Thus, using the mean value theorem for analytic functions,

$$\left\langle u, \frac{g(0)}{W_2(0)}\right\rangle = \int_0^{2\pi} \left\langle u, \frac{g(e^{i\theta})}{W_2(e^{i\theta})}\right\rangle \frac{d\theta}{2\pi} = L_v(F_0) = \frac{\langle u,v\rangle}{W_2(0)}.$$

Thus, $\langle u,g(0)\rangle = \langle u,v\rangle$ for all $u \in \mathbb{C}^n$; therefore, $g(0) = v$. If $N_0(v)$ denotes the intermediate norm of $v$ obtained, by the method of theorem I, from the boundary norms $|\ |^*_{e^{i\theta}}$ we then must have $N_0(v) \leq \|\!|g|\!\|^*_2 = \|L_v\| \leq |v|^*_0$. That is,

$$N_0(v) \leq |v|^*_0. \tag{3.15}$$

We shall now show that the reverse inequality also holds. This, then, would establish theorem II. Consider two extremal functions $F = F_{0,w}$ and $G = G_{0,v}$, the first corresponding to the boundary norms $\{|\ |_{e^{i\theta}}\}$, the second corresponding to $\{|\ |^*_{e^{i\theta}}\}$. Thus, by theorem I, $N_0(v) = \|\!|G|\!\|^*_2$; also $W_1F$ and $W_2^{-1}G$ belong to $H^2(D;\mathbb{C}^n)$. Hence, $(W_1/W_2)\langle F,G\rangle \in H^1(D)$. By the subharmonicity of the log of the modulus of this function we have

$$\log\left|\frac{W_1(0)}{W_2(0)}\langle F(0),G(0)\rangle\right| \leq \int_0^{2\pi} \log\left|\frac{W_1(e^{i\theta})}{W_2(e^{i\theta})}\langle F(e^{i\theta}),G(e^{i\theta})\rangle\right| \frac{d\theta}{2\pi}.$$

But, $\log\left|\frac{W_1(0)}{W_2(0)}\right| = \int_0^{2\pi} \log\left|\frac{W_1(e^{i\theta})}{W_2(e^{i\theta})}\right| \frac{d\theta}{2\pi}$. Consequently, since F and G are extremal,

$$\log |\langle w,v\rangle| \leq \int_0^{2\pi} \log|\langle F(e^{i\theta}),G(e^{i\theta})\rangle| \frac{d\theta}{2\pi}$$

$$\leq \int_0^{2\pi} \log|F(e^{i\theta})|_{e^{i\theta}} |G(e^{i\theta})|^*_{e^{i\theta}} \frac{d\theta}{2\pi} = \log|w|_0 N_0(v).$$

That is, $|\langle w,v\rangle| \leq |w|_0 N_0(v)$. If we now consider the supremum over all $w \in \mathbb{C}^n$ with $|w|_0 = 1$ we then have the desired reverse inequality: $|v|_0^* \leq N_0(v)$.

## §4. Basic properties of the intermediate spaces

Theorems I and II have many consequences. We shall now present some of them and illustrate the importance of the duality result we just obtained.

If we apply inequality (3.9) to the dual norms obtained as intermediate spaces of the family $\{| \ |^*_{e^{i\theta}}\}$ we obtain

$$|v|^*_{z_0} \leq \frac{1}{|W_1(z_0)|} \|v\|.$$

Using (3.10) and (3.11), together with (3.9), we obtain

Corollary (4.1). $|W_1(z_0)|\|v\| \leq |v|_{z_0} \leq |W_2(z_0)|\|v\|$ for all $z_0 \in D$ and $v \in \mathbb{C}^n$.

This tells us that if $\Omega$ is a domain whose closure is in $D$ we have the condition corresponding to (3.1) on the boundary $\partial\Omega$ which we needed for the construction of theorem I. That is, assuming $\Omega$ has a Herglotz kernel associated with it, the spaces $B_\zeta = (\mathbb{C}^n, | \ |_\zeta)$, $\zeta \in \partial\Omega$, can be used to obtain intermediate spaces $[B_\zeta]_z$ for all $z \in \Omega$. It is natural to inquire what are the relations between these spaces and $B_z$:

Corollary (4.2) (The iteration theorem.) If $\overline{\Omega} \subset D$ is a subdomain of the type just described then $[B_\zeta]_z = B_z$ for all $z \in \Omega$.

Proof: Suppose $N_{z_0}$ is the norm of $[B_\zeta]_{z_0}$, $\zeta \in \partial\Omega$. If $F \in H^\infty_\#(D)$ then, by definition, $|F(z)|_z \leq |||F|||_\infty$ for all $z \in D$. Thus,

$$|v|_{z_0} \geq \inf\{\sup_{\zeta\in D} |F(\zeta)|_\zeta : F \in H^\infty_\#(D),\ F(z_0) = v\}$$

$$\geq \inf\{\operatorname{ess.\,sup}_{\zeta\in\partial\Omega} |F(\zeta)|_\zeta : F \in H^\infty_\#(\Omega),\ F(z_0) = v\} = N_{z_0}(v).$$

Using theorem II and this last result, we then have $|v|^*_{z_0} \geq N^*_{z_0}(v)$. Taking the dual norms of both sides of this inequality and using (3.10) we then have $|v|_{z_0} \leq N_{z_0}(v)$. This proves (4.2).

We have seen (Corollary (3.8)) that, if $F$ is an extremal function associated with $z_0 \in D$ and $v \in \mathbb{C}^n$, then $|F(e^{i\theta})|_{e^{i\theta}} = |v|_{z_0}$ a.e. It turns out that in many cases the extremal function corresponding to $z_0$ and $v$ is unique. This fact is also a corollary of Theorem II. The property guaranteeing this uniqueness is usually referred to as <u>smoothness</u>: $B = (\mathbb{C}^n, N)$ is <u>smooth</u> provided for each $v \in \mathbb{C}^n$ there exists a unique $w \in \mathbb{C}^n$ such that $N^*(w) = 1$ and $\langle v,w\rangle = N(v)$ (for the connection between this notion and the uniform convexity of the dual of $B$ see [4]).

<u>Corollary (4.3)</u>. If the boundary spaces are smooth, then the extremal function, $F_{z_0,v}$, associated with a point $z_0$ in the domain and a vector $v \in \mathbb{C}^n$ is unique.

<u>Proof</u>: By corollary (3.8) an extremal function $F$ associated with $z_0$ and $v$ must belong to $H^\infty_\#$, satisfies $|F(e^{i\theta})|_{e^{i\theta}} = |v|_{z_0}$ a.e. and $F(z_0) = v$. Let us choose $w \in \mathbb{C}^n$ with $|w|^*_{z_0} = 1$ and satisfying $\langle v,w\rangle = |v|_{z_0}$. Suppose $G$ is an extremal function corresponding to $z_0$ and $w$ obtained from the dual norms $\{|\ |^*_{e^{i\theta}}\}$ on the boundary of $D$. Thus, in particular, $|G(e^{i\theta})|^*_{e^{i\theta}} = 1$ a.e. Then the function $\langle F(z),G(z)\rangle$ belongs to $H^\infty(D)$ (since $|\langle F(z),G(z)\rangle| \leq |F(z)|_z |G(z)|^*_z \leq \|\|F\|\|_\infty \|\|G\|\|^*_\infty = |v|_{z_0}$). Consequently,

$$|v|_{z_0} = \langle v,w\rangle = \langle F(z_0),G(z_0)\rangle = \int_0^{2\pi} P_{z_0}(\theta)\langle F(e^{i\theta}),G(e^{i\theta})\rangle\, d\theta$$

$$\leq \int_0^{2\pi} \|\|F\|\|_\infty \|\|G\|\|^*_\infty P_{z_0}(\theta)\, d\theta = |v|_{z_0}.$$

Since $|\langle F(e^{i\theta}),G(e^{i\theta})\rangle| \leq |v|_{z_0}$ a.e., the equality (which, by the above, must hold)

$$\int_0^{2\pi} P_{z_0}(\theta)\langle F(e^{i\theta}),G(e^{i\theta})\rangle\, d\theta = |v|_{z_0}$$

implies $\langle F(e^{i\theta}),G(e^{i\theta})\rangle = |v|_{z_0} = |F(e^{i\theta})|_{e^{i\theta}}$ a.e. Since

$|G(e^{i\theta})|^*_{e^{i\theta}} = 1$ a.e., the smoothness assumption tells us that $F(e^{i\theta})$ is a.e. determined by $G$. But this uniquely determines the analytic function $F$.

If $F$ is analytic in a domain $D$ then, as is well known, $\log|F|$ is a subharmonic function. We shall show that this subharmonicty property holds when intermediate norms are used. More precisely, suppose we have the situation of theorem I and $F$ is a $\mathbb{C}^n$-valued analytic function on $D$. Let $z_0 \in D$ and suppose the closed disc about $z_0$ of radius $r > 0$ is contained in $D$. If $w(\theta) = z_0 + re^{i\theta}$ then, by definition,

$$|F(z_0)|_{z_0} \leq \left(\int_0^{2\pi} |F(w(\theta))|^p_{w(\theta)} \frac{d\theta}{2\pi}\right)^{1/p}$$

for all $p > 0$ (the integral is well defined since $|F(w(\theta))|_{w(\theta)} \leq |W_2(w(\theta))| \, \|F(w(\theta))\|$ is bounded for $\theta \in [0,2\pi)$). Thus, letting $p \to 0$ the right side of the above inequality tends to

$$\exp\left\{\int_0^{2\pi} \log|F(w(\theta))|_{w(\theta)} \frac{d\theta}{2\pi}\right\}.$$

By taking logarithms this shows (under the hypotheses of theorem I):

Corollary (4.4). If $F$ is a $\mathbb{C}^n$-valued analytic function on $D$ then $\log|F(z)|_z$ is subharmonic.

This result, among other things, is useful for obtaining a characterization of extremal functions. Suppose $F = F_{z_0,v}$ is an extremal function; then for any $z \in D$

$$|F(z)|_z \leq |||F|||_\infty = |v|_{z_0} = |F(z_0)|_{z_0}.$$

But, by the maximum principal for subharmonic functions, this means that $|F(z)|_z$ is constant. It is this constancy property that characterizes extremal functions:

Corollary (4.5). If $F$ is an extremal function corresponding to $z_0 \in D$ and $v \in \mathbb{C}^n$ then $|F(z)|_z = |v|_{z_0}$ for all $z \in D$. Conversely, if $F$ is an analytic $\mathbb{C}^n$-valued function such that $|F(z)|_z = c$ for all $z \in D$ then it is an extremal function for each $z \in D$ and vector $F(z)$.

Proof: Suppose the analytic function $F$ satisfies $|F(z)|_z = c$ for all $z \in D$. Then $\|W_1(z)F(z)\| \le |F(z)|_z = c$ for all $z \in D$, by corollary (4.1). Thus, $W_1F \in H^\infty(D;\mathbb{C}^n)$. We claim that $F \in H^\infty_\#$ and $c = |||F|||_\infty$. This would imply that $F$ is an extremal function associated with $z$ and $F(z)$ for each $z \in D$ and the corollary would be established.

To see this, choose $w \in \mathbb{C}^n$, let $W(z) = \exp \int_0^{2\pi} h_z(\theta)\log|w|^*_{e^{i\theta}}d\theta$ and put $G(z) = W(z)^{-1}w$. Since $1/|w|^*_{e^{i\theta}} \le k_2(\theta)/\|w\|$ (by (3.11)) we have

$$\|G(z)\| \le \|w\| \exp \int_0^{2\pi} P_z(\theta) \log\left(\frac{k_2(\theta)}{\|w\|}\right) d\theta = |W_2(z)|.$$

Thus, $\left\|\frac{1}{W_2(z)} G(z)\right\| \le 1$ and, therefore, $G \in H_1^{\infty *}$. Since $G(e^{i\theta}) = \frac{w}{W(e^{i\theta})}$ we have

$$|||G|||^*_\infty = \operatorname*{ess.\,sup}_{\theta\in[0,2\pi)} |G(e^{i\theta})|^*_{e^{i\theta}} = \operatorname*{ess.\,sup}_{\theta\in[0,2\pi)} \frac{|w|^*_{e^{i\theta}}}{|w|^*_{e^{i\theta}}} = 1.$$

Hence, $|\langle F(z),G(z)\rangle| \le |F(z)|_z|G(z)|^*_z \le c\cdot 1 = c$, which implies $\langle F(z),G(z)\rangle$ belongs to $H^\infty(D)$. It has boundary values $\langle F(e^{i\theta}),G(e^{i\theta})\rangle = \left\langle F(e^{i\theta}), \frac{w}{W(e^{i\theta})}\right\rangle$ a.e. It follows that

$$|\langle F(e^{i\theta}),w\rangle| \le c|W(e^{i\theta})| = c|w|^*_{e^{i\theta}} \quad \text{a.e.}$$

for all $w$. From this we see that $|F(e^{i\theta})|_{e^{i\theta}} \le c$. But $c = |F(z)|_z \le |||F|||_\infty \le c$; thus, $|F(e^{i\theta})|_{e^{i\theta}} = c$ a.e. and the corollary is proved.

Let us now suppose our boundary Banach spaces satisfy the smoothness condition giving us the uniqueness of the extremal functions (see Corollary (4.3)). We introduce the notation

$$A(z,z_0)(v) = F_{z_0,v}(z)$$

for $z,z_0 \in D$ and $v \in \mathbb{C}^n$. That is, for each $z_0 \in D$ we have a mapping $A(z,z_0)\colon \mathbb{C}^n \to \mathbb{C}^n$ such that $A(\cdot,z_0)(v)$ is the extremal function assciated with $z_0$ and $v$. In particular, $A(z,z_0)(v)$ is an analytic function of $z \in D$. In general, $A(z,z_0)$ is not linear; however, it does satisfy many basic properties. We shall now derive some of them. The following is an immediate consequence of the definition of extremal functions:

(4.6) $A(z,z) = I =$ identity operator, for all $z \in D$.

Let us choose $z_0, z_1 \in D$ and $v \in \mathbb{C}^n$ and put $F(z) = A(z,z_0)[A(z_0,z_1)(v)]$ for $z \in D$. Then $F$ is analytic and is the extremal function associated with $z_0$ and $A(z_0,z_1)(v)$. The function $G(z) = A(z,z_1)(v)$ is also an extremal function. In fact, by (4.5), it is the unique extremal function associated with $z$ and $G(z)$. But $G(z_0) = A(z_0,z_1)(v) = F(z_0)$. Thus, $F(z) \equiv G(z)$. This shows:

(4.7) (The propagator equation). $A(z,z_0) \circ A(z_0,z_1) = A(z,z_1)$ for all $z, z_0, z_1 \in D$.

From these two properties we obtain

(4.8) $A(z,z_0)$ maps $\mathbb{C}^n$ one-to-one onto $\mathbb{C}^n$ and $A(z_0,z)$ is both the left and right inverse of $A(z,z_0)$.

Let $A(z) = A(z,0)$ and $a(z) = A(0,z)$. We then have, from (4.7) and the constancy property (4.5):

(4.9) $$A(z,z_0) = A(x) \circ a(z_0), \qquad |v|_z = |a(z)(v)|_0.$$

The last equality tells us that the mapping $a(z): B_z = (\mathbb{C}^n, |\ |_z) \to B_0 = (\mathbb{C}^n, |\ |_0)$ is norm preserving (more generally, this is true of $A(z,z_0): B_{z_0} \to B_z$). Since $|\ |_z$ is a norm we have the subadditivity property:

(4.10) $$|a(z)(v_1+v_2)|_0 \leq |a(z)(v_1)|_0 + |a(z)(v_2)|_0.$$

The uniqueness of the extremal functions gives us the homogeneity property:

(4.11) $A(z,z_0)(\lambda v) = \lambda A(z,z_0)$ for $\lambda \in \mathbb{C}$, $z, z_0 \in D$ and $v \in \mathbb{C}^n$.

(4.12) $A(z,z_0): \mathbb{C}^n \to \mathbb{C}^n$ is continuous for each pair $z, z_0 \in D$.

Proof: Choose $v \in \mathbb{C}^n$ and a sequence $\{v_k\}$ converging to $v$. Let $F_k(z) = A(z,z_1)(v_k)$. Then $|v_k|_{z_0} = |||F_k|||_{2,z_0}$ and

$\lim_{k\to\infty} |v_k|_{z_0} = |v|_{z_0}$. Thus, $\{F_k\}$ is a bounded sequence in $H^2_{\#} \subset L^2_{\#}$. We saw in the proof of theorem II (see footnote (3)) that $L^2_{\#}$ is reflexive; therefore, a subsequence of $\{F_k\}$ converges weakly to an element $F \in L^2_{\#}$. But $H^2_{\#}$ is a convex set which is closed in the norm topology of $L^2_{\#}$; thus, $H^2_{\#}$ is also a weakly closed set. Consequently $F \in H^2_{\#}$. Relabelling, if necessary, we can assume $\{F_k\}$ itself converges to $F$ weakly.

Choose $w \in \mathbb{C}^n$ and $z \in D$. The linear functional mapping $G \in H^2_{\#}$ into $\langle G(z),w\rangle$ is continuous. Thus, $\lim_{k\to\infty} \langle F_k(z),w\rangle = \langle F(z),w\rangle$. Two immediate consequences of this convergence is that $F(z_0) = v$ and $|F(z)|_z = \lim_{k\to\infty} |F_k(z)|_z = \lim_{k\to\infty} |v_k|_{z_0} = |v|_{z_0}$. Thus, $F$ has the constancy property

$$|F(z)|_z = |v|_{z_0}, \qquad z \in D,$$

and, by corollary (4.5), it must be the extremal function $F(z) = A(z,z_0)(v)$. But the convergence $\lim_{k\to\infty}\langle F_k(z),w\rangle = \langle F(z),w\rangle$, for each $w \in \mathbb{C}^n$, also implies $\lim_{k\to\infty} \|F_k(z)-F(z)\| = \lim_{k\to\infty} \|A(z,z_0)(v_k)-A(z,z_0)(v)\| = 0$.

We have shown, therefore, that whenever $v_k \to v$ as $k \to \infty$ then there exists a subsequence $\{v_{k_j}\}$ such that

$$\lim_{j\to\infty} \|A(z,z_0)(v_{k_j}) - A(z,z_0)(v)\| = 0$$

But this clearly implies $\lim_{k\to\infty} \|A(z,z_0)(v_k) - A(z,z_0)(v)\| = 0$ and the desired continuity is proved.

(4.13) $$\lim_{z \rhd e^{i\theta}} |v|_z = |v|_{e^{i\theta}} \quad \text{a.e.}$$

<u>Proof</u>: Let us assume that the duals of the boundary space are also smooth so that we have the uniqueness of the corresponding extremal functions $A^*(z,z_0)(w)$. We write $A^*(z) = A^*(z,0)$. Select a countable dense subset, $\{w_j \in \mathbb{C}^n : |w_j|_0^* = 1,\ j = 1,2.3,\dots\}$, of the surface of the unit sphere in $\mathbb{C}^n$. Then for almost every $\theta$ we have

(a) $\lim_{z \rhd e^{i\theta}} A^*(z)(w_j) = A^*(e^{i\theta})(w_j)$,

(b) $|A^*(e^{i\theta})(w_j)|_{e^{i\theta}} = 1$,

(c) $\lim_{z \rhd e^{i\theta}} |W_2(z)| = k_2(\theta) \neq 0.$

Let us fix $v \in \mathbb{C}^n$. Then

(d) $|v|_z = \sup_{j=1,2,\ldots} |<v,A^*(z)(w_j)>|.$

This is an immediate consequence of the fact that $|A^*(z)|_z^* = |w|_0^*$, the density of $\{w_j\}$ in the surface of the unit sphere of $\mathbb{C}^n$, the continuity of $A^*(z): \mathbb{C}^n \to \mathbb{C}^n$ ((4.12)) and the onto property of this map ((4.8)). Let $f_j(z) = |<v,A^*(z)(w_j)>| / |W_2(z)|$. Then $f_j$ is the absolute value of an analytic function in $H^\infty(D)$ since, using (4.1),

$$f_j(z) \leq |v|_z |A^*(z)(w_j)|_z^* / |W_2(z)| = |v|_z / |W_2(z)| \leq \|v\|.$$

Thus, by (a) and (c), $f_j(e^{i\theta}) = \lim_{z \rhd e^{i\theta}} f_j(z) = |<v,A^*(e^{i\theta})(w_j)>|/k_2(\theta)$ a.e.; but, by (b),

$$|<v,A^*(e^{i\theta})(w_j)>| \leq |v|_{e^{i\theta}} |A^*(e^{i\theta})(w_j)|^*_{e^{i\theta}} = |v|_{e^{i\theta}}.$$

Hence,

(e) $f_j(e^{i\theta}) \leq |v|_{e^{i\theta}}/k_2(\theta) \leq \|v\|.$

Let $h(e^{i\theta}) = \sup_j f_j(e^{i\theta})$. By the last inequality $h(e^{i\theta}) \leq \|v\|$. Thus, the Poisson integral of $h$ is defined and $h(z) \geq \int_0^{2\pi} P_z(\theta) f_j(e^{i\theta})\, d\theta$. But, from (d), we have $\sup_j f_j(z) = |v|_z / |W_2(z)|$. Consequently, $|v|_z \leq |W_2(z)| h(z)$. But, by (e),

$$|v|_{e^{i\theta}} \geq k_2(\theta) h(e^{i\theta}) = \lim_{z \rhd e^{i\theta}} |W_2(z)| h(z) \geq \overline{\lim_{z \rhd e^{i\theta}}} |v|_z, \quad \text{a.e.}$$

By theorem II, we also have

(f) $|v|^*_{e^{i\theta}} \geq \overline{\lim_{z \rhd e^{i\theta}}} |v|_z^*$ a.e.

Let us choose $w \in \mathbb{C}^n$. We have just shown $|w|_{e^{i\theta}} \geq \overline{\lim_{z \rhd e^{i\theta}}} |w|_z$ a.e. We shall now use (f) to obtain $|w|_{e^{i\theta}} \leq \underline{\lim_{z \rhd e^{i\theta}}} |w|_z$ a.e. Let $e^{i\theta}$ be a point on the boundary of $D$ and denote by $\Omega(\theta)$ a "pointer" region with vertex at $e^{i\theta}$; let $\Omega_\varepsilon(\theta) = \{z \in \Omega(\theta) : |z| > 1-\varepsilon\}$. (See figure.):

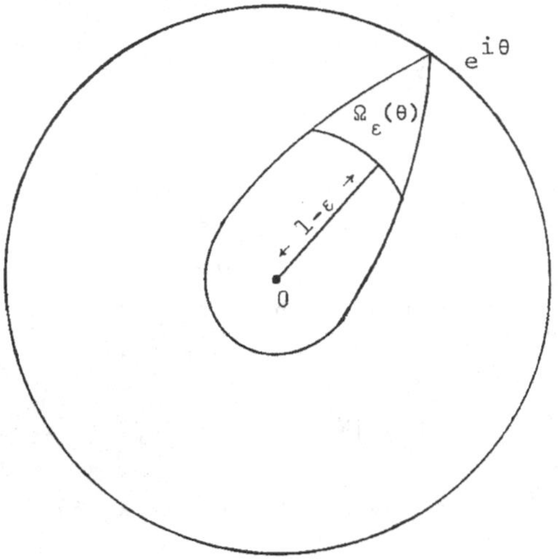


Choose a countable dense set $\{v_j\}$ in the surface of the unit sphere of $\mathbb{C}^n$. We then have (f) holding for $v = v_j$, $j = 1,2,\ldots,$ in a set $E \subset [0,2\pi)$ of measure $2\pi$. Let $\theta \in E$ and $1 > \delta > 0$. We can then find $v = \dfrac{v_j}{|v_j|^*_{e^{i\theta}}}$ such that $|\langle w,v\rangle| \geq (1-\delta)|w|_{e^{i\theta}}$. For this $v$ and $\theta$ we can find, using (f), $\varepsilon > 0$ such that $(1+\delta) = |v|^*_{e^{i\theta}}(1+\delta) \geq |v|^*_z$ for all $z \in \Omega_\varepsilon(\theta)$. Thus,

$$(1-\delta)|w|_{e^{i\theta}} \leq |\langle w,v\rangle| \leq |w|_z|v|^*_z \leq |w|_z(1+\delta).$$

This shows that $|w|_{e^{i\theta}} \leq \frac{1+\delta}{1-\delta}|w|_z$ for all $z \in \Omega_\varepsilon(\theta)$. Since $\delta > 0$ can be chosen arbitrarily small it follows that $|w|_{e^{i\theta}} \leq \underline{\lim}_{z \rhd e^{i\theta}} |w|_z$. This proves (4.12).

Let us now pass to the problem of interpolation of operators mapping a family of intermediate spaces satisfying these properties into another such family. We shall see that theorem (2.1) has a natural extension to this situation. Suppose $\{B_z\} = \{(\mathbb{C}^n,N_z)\}$ and $\{C_z\} = \{(\mathbb{C}^n,M_z)\}$ are two such families and that $T_z\colon B_z \to C_z$ are linear operators. We assume $\{T_z\}$ is an <u>analytic family</u>; we mean by this that $z \to T_z v$ is an analytic $\mathbb{C}^n$-valued function of $z \in D$. We choose a point $z_0 \in D$ and vectors $v,w \in \mathbb{C}^n$ such that $N_{z_0}(v) = 1 = M^*_{z_0}(w)$. Let $F = F_{z_0,v}$ be an extremal function associated with $z_0,v$ and the norms $\{N_z\}$ and $G = G_{z_0 w}$ an extremal function associated with $z_0$, $w$ and the dual norms $\{M^*_Z\}$. It is easily checked that

$$\Phi(z) = \langle T_z F(z),G(z)\rangle$$

is an analytic function on $D$ (we can see this most easily by using (4.1) to estimate $\Phi(z)$ in terms of Euclidean norms: if $K \subset D$ is compact there exists a constant $c_K$ such that

$$\mathfrak{N}(z) \le c_K \sup_{\|v\|\le 1} \|T_z v\|, \qquad z \in K.$$

Then using this inequality, again (4.1), and the fact that $W_1 F \in H^\infty(D;\mathbb{C}^n)$ we can deduce that $T_z F(z)$ is an analytic $\mathbb{C}^n$-valued function of $z \in D$).

Suppose the closed disc of radius $\rho > 0$ about $z_0$ is contained in $D$. Writing $\zeta(\theta) = z_0 + \rho e^{i\theta}$, $\theta \in [0,2\pi)$, using the logarithmic subharmonicity of $|\Phi(z)|$ and Corollary (4.5), we have

$$\begin{aligned}
\log|\Phi(z_0)| &\le \frac{1}{2\pi}\int_0^{2\pi} \log|\Phi(\zeta(\theta))|\, d\theta \\
&\le \frac{1}{2\pi}\int_0^{2\pi} \log \mathfrak{N}(\zeta(\theta) N_{\zeta(\theta)}(F(\zeta(\theta))M^*_{\zeta(\theta)}(G(\zeta(\theta))\, d\theta \\
&= \frac{1}{2\pi}\int_0^{2\pi} \log \mathfrak{N}(\zeta(\theta))\, d\theta\; N_{z_0}(v)M^*_{z_0}(w) \\
&= \frac{1}{2\pi}\int_0^{2\pi} \log \mathfrak{N}(\zeta(\theta))\, d\theta .
\end{aligned}$$

Now, taking the supremum over all $v$, $w$ satisfying $N_{z_0}(v) = 1 = M^*_{z_0}(w)$ we obtain

Theorem (4.14) (The Interpolation Theorem). If $\{T_z\}$ is an analytic family of linear operators mapping $B_z$ into $C_z$ for $z \in D$ and $\mathfrak{N}(z)$ is the operator norm of $T_z$, then $\log \mathfrak{N}(z)$ is a subharmonic function.

As is the case in theorem (2.1) one can apply this result when we are given information about $\mathfrak{N}(\zeta)$, $\zeta \in \partial D$, by constructing the intermediate spaces from the corresponding boundary spaces $\{B_\zeta\}$ and $\{C_\zeta\}$ by the method of theorem I. In such applications we have to check that the analytic family $\{T_z\}$ satisfies "admissible growth" conditions of the kind mentioned in §2.

Before giving some applications of these results let us make an observation concerning our construction of intermediate spaces. We have solved a general version of the Dirichlet problem for the region $D$ when the boundary data is not necessarily a numerical valued function but is a function whose value at $\zeta \in \partial D$ is the norm $|\ |_\zeta$ (or the Banach space $B_\zeta = (\mathbb{C}^n, |\ |_\zeta)$). If $n = 1$ then, by homogeneity this boundary norm has the form $|v|_\zeta = k(\zeta)\|v\|$ and the basic assumption (3.1) can be made with $k_1 = k_2 = k$, where $\log k$ is integrable. Thus, $W_1(z) = W_2(z) \equiv W(z)$ and, from (4.1) we see that $|v|_{z_0} = |W(z_0)|\,\|v\|$. We see, therefore, that what we have done is used the solution of the Dirichlet problem with the boundary function $\log k$ and the intermediate norms are obtained by multiplying the Euclidean norm (here, it's the absolute value) by the exponential of this solution. For these reasons we call the family of intermediate spaces $\{B_z\}$ we constructed a log-harmonic family (which is determined by the boundary data $\{B_\zeta\}$, $\zeta \in \partial D$). The basic property characterizing these families is the Iteration Theorem (4.2) which plays the role that the mean value theorem does for harmonic functions. A more complete discussion of these notions is given in [3].

## §5. Some applications

In the second section we considered a family of $L^p$-spaces associated with the points of a domain $D$. They were obtained from "boundary" $L^{p(\zeta)}$-spaces by solving the Dirichlet problem for the function $\frac{1}{p(\zeta)}$. We claim that these are the same spaces that would have been obtained by the method of theorem I. For simplicity suppose $D$ is the unit disc and we are given a family of $L^{p(\theta)}$-norms on $C^n$:

$$(5.1) \qquad |v|_{e^{i\theta}} = \|v\|_{p(\theta)} = \Big(\sum_{k=1}^{n} |v_k|^{p(\theta)} m_k\Big)^{1/p(\theta)},$$

where $v = (v_1, v_2, \ldots, v_n) \in \mathbb{C}^n$, $0 < m_k$ and $p(\theta)$ is a measurable function such that $1 \le p(\theta) \le \infty$. Then, clearly, there exist constants $k_1$ and $k_2$ such that

$$(5.2) \qquad k_1\|v\| \le |v|_{e^{i\theta}} \le k_2\|v\|;$$

consequently, condition (3.1) is certainly satisfied. In view of (2.3) and (4.5) it is reasonable to expect the function defined immediately before (2.3) to be an extremal function. That is, if $v \in \mathbb{C}^n$ and $z_0 \in D$ the function of $z$, $B(z,z_0)(v)$, whose $k^{th}$ coordinate is

$$v_k\left\{\frac{|v_k|}{\|v\|_{p(z_0)}}\right\}^{[\alpha(z)-\alpha(z_0)]p(z_0)},$$

where $\alpha(z) = \int_0^{2\pi} h_z(\theta)\,\frac{d\theta}{p(e^{i\theta})}$, should, if there were justice in this world, be an extremal function associated with $v$ and $z_0$. It turns out that, at least in this case, there is justice. First, observe that

$B(z,z_0)(v)$ is analytic in $z$ and $B(z_0,z_0)(v) = v$. Moreover, if $|z| \le 1$ we have, analogously to (2.3),

$$\|B(z,z_0)(v)\|_{p(z)} = \|v\|_{p(z_0)} \tag{5.3}$$

for all $z \in D$ (this is a straightforward computation). From this we see immediately that $B(\cdot,z_0)(v) \in H^\infty_\#$ and, thus

$$|v|_{z_0} \le \|v\|_{p(z_0)}. \tag{5.4}$$

We shall now show that the reverse inequality is also true. Let $A(z)$ be an extremal function associated with $z_0$ and $v$. With $\frac{1}{q(z)} = 1 - \frac{1}{p(z)}$ choose $w \in \mathbb{C}^n$ such that $\|w\|_{q(z_0)} = 1$ and denote by $\tilde{B}(z,z_0)(w)$ the function just defined above in terms of $q(z)$ (instead of $p(z)$) and $w$ (instead of $v$). Then, by (5.4) and (5.2), $|\tilde{B}(z,z_0)(w)|^*_z \le \|B(z,z_0)(w)\|_{q(z)} = \|w\|_{q(z_0)}$. Thus

$$\begin{aligned} |\langle A(z),\tilde{B}(z,z_0)(w)\rangle| &\le |A(z)|_z|\tilde{B}(z,z_0)(w)|^*_z = |v|_{z_0}|\tilde{B}(z,z_0)(w)|^*_z \\ &\le |v|_{z_0}\|w\|_{q(z_0)} = |v|_{z_0}. \end{aligned}$$

Consequently $\Phi(z) = \langle A(z),\tilde{B}(z,z_0)(w)\rangle$ belongs to $H^\infty(D)$ and, therefore,

$$\langle v,w\rangle = \Phi(z_0) = \int_0^{2\pi} P_{z_0}(\theta)\Phi(e^{i\theta})\,d\theta.$$

Hence,

$$\begin{aligned} |\langle v,w\rangle| &\le \int_0^{2\pi} P_{z_0}(\theta)\|A(e^{i\theta})\|_{p(e^{i\theta})}\|\tilde{B}(e^{i\theta},z_0)(w)\|_{q(e^{i\theta})}\,d\theta \\ &= \left(\int_0^{2\pi} P_{z_0}(\theta)\,d\theta\right)|v|_{z_0}\|w\|_{q(z_0)} = |v|_{z_0}. \end{aligned}$$

If we now take the supremum over all $w$ satisfying $\|w\|_{q(z_0)} = 1$ we see that $\|v\|_{p(z_0)} \le |v|_{z_0}$. This shows (because of (5.4)) that

$$|v|_{z_0} = \|v\|_{p(z_0)}$$

for all $z_0 \in D$. The claim made at the beginning of this section is, therefore, established.

Let us now turn to another interesting class of intermediate spaces obtained when each of the boundary norms is a Hilbert space norm: for each $\theta \in [0,2\pi)$ let $\Gamma(\theta)$ be an $n \times n$ invertible matrix and, for $v \in \mathbb{C}^n$, let $|v|_{e^{i\theta}} = \|\Gamma(\theta)v\|$. We assume $\theta \to \Gamma(\theta)$ is a measurable function. Then, clearly, letting $\|\Gamma(\theta)\|$ and $\|\Gamma(\theta)^{-1}\|$ denote operator norms,

$$(5.5) \qquad \frac{1}{\|\Gamma(\theta)^{-1}\|}\|v\| \leq |v|_{e^{i\theta}} \leq \|\Gamma(\theta)\|\|v\|.$$

Then, the condition (3.1) is satisfied if we assume $\log \|\Gamma(\theta)\|$ and $\log \|\Gamma(\theta)^{-1}\|$ are integrable. By theorem I we have

$$|v|_{z_0} = \inf\{|||F|||_{2,z_0} : F \in H^2_{\#},\ F(z_0) = v\}$$

whenever $z_0 \in D$ and $v \in \mathbb{C}^n$. In this situation we can express the extremal function $A(z,z_0)(v)$ in terms of a projection operator acting on the Hilbert space $H^2_{\#}$ endowed with the norm $|||\ |||_{2,z_0}$. Let $H_0 = \{F \in H^2_{\#} : F(z_0) = 0\}$ and $P$ be the projection of $H^2_{\#}$ onto $H_0^{\perp}$. The function $G^v_{z_0} = \frac{W_2(z_0)}{W_2(z)} v$ belongs to $H^2_{\#}$ (see the beginning of the proof of theorem I). Then, clearly, the set $E = E^v_{2,z_0}$ introduced in the proof of theorem I is the coset $G^v_{z_0} + H_0$.

<u>Lemma (5.6).</u> $(PG^v_{z_0})(z) = A(z,z_0)(v)$ for all $z_0$, $v$.

<u>Proof</u>: Since the boundary spaces are smooth we know by (4.3) that the extremal function associated with $z_0$ and $v$ is unique. Thus, all we need to do is check that, among all members of the coset $G^v_{z_0} + H_0$, $PG^v_{z_0}$ has minimal norm (obviously $G^v_{z_0}(z_0) = v$). Suppose $G^v_{z_0} = h + h^{\perp}$ with $h \in H_0$ and $h^{\perp} \in H_0^{\perp}$. The general element of $G^v_{z_0} + H_0$, therefore, has the form $(h+h_0) + h^{\perp}$ with $h_0 \in H_0$. Consequently,

$$|||(h+h_0) + h^{\perp}|||^2_{2,z_0} = |||h+h_0|||^2_{2,z_0} + |||h^{\perp}|||^2_{2,z_0} \geq |||h^{\perp}|||^2_{2,z_0}$$
$$= |||PG^v_{z_0}|||^2_{2,z_0}.$$

This proves the lemma.

The mapping $v \to (PG^v_{z_0})(z)$ is clearly linear since $P$ is a linear operator on $H^2_{\#}$. Thus, $A(z,z_0)$ is a linear operator on $\mathbb{C}^n$ which is analytic in $z$. By Cramer's rule, the inverse, $A(z_0,z)$, must also be analytic in $z$. This is certainly not true in general (see the form exhibited after (5.2) for the operator $B(z,z_0)$ when the boundary spaces were $L^p$-spaces). Thus, we can consider $A(z,z_0)$ to be an $n \times n$ matrix valued function which is analytic in either of the two variables $z$ and $z_0$. From (4.9) we have, in fact, that $A(z,z_0)$ is the product $A(z)a(z_0)$; each of the factors, $A(z)$ and $a(z_0)$, is analytic. Moreover, $|v|_z = |a(z)v|_0$. By (4.1) we have

$$\left|\frac{a(z)}{W_2(z)}\, v\right|_0 \le \|v\| .$$

Thus, $\frac{a(z)}{W_2(z)}$ is a bounded, matrix-valued analytic function. Hence we have the almost everywhere existence of the non-tangential limits

$$\lim_{z \rhd e^{i\theta}} \frac{a(z)}{W_2(z)} = \frac{a(e^{i\theta})}{W_2(e^{i\theta})} . \tag{5.7}$$

But

$$|v|_0 = \left(\int_0^{2\pi} |A(e^{i\theta})(v)|^2_{e^{i\theta}} \frac{d\theta}{2\pi}\right)^{1/2} = \left(\int_0^{2\pi} \|\Gamma(\theta)A(e^{i\theta})(v)\|^2 \frac{d\theta}{2\pi}\right)^{1/2}$$

is a Hilbert space norm on $\mathbb{C}^n$. Thus, there exists a matrix $\alpha$ such that $|v|_0 = \|\alpha v\|$ for all $v \in \mathbb{C}^n$. We must have, therefore, because of (5.7),

$$\lim_{z \rhd e^{i\theta}} |v|_z = \lim_{z \rhd e^{i\theta}} \|\alpha a(z)v\| = \|\alpha a(e^{i\theta})v\|$$

a.e. On the other hand, by (4.13),

$$\lim_{z \rhd e^{i\theta}} |v|_z = |v|_{e^{i\theta}} = \|\Gamma(\theta)v\| \quad \text{a.e.}$$

This shows that $\|\alpha a(e^{i\theta})v\| = \|\Gamma(\theta)v\|$ a.e. If we let $b(z) = \alpha a(z)$ for $z \in \overline{D}$ and put $P(\theta) = \Gamma(\theta)^*\Gamma(\theta)$ we obtain the following result:

<u>Theorem (5.8)</u>. Suppose $P(\theta)$ is a positive definite matrix for each $\theta \in [0,2\pi)$ such that $\log\|P(\theta)\|$ and $\log\|P(\theta)^{-1}\|$ are integrable, then there exists an analytic matrix valued function $b(z)$ on $D$ such that

$$\lim_{z \rhd e^{i\theta}} b(z)^* b(z) = P(\theta) \tag{5.9}$$

almost everywhere. Moreover, the operator norm $\|b(z)\|$ satisfies

$$(5.10) \qquad k_1(z) \le \|b(z)\| \le k_2(z)$$

for all $z \in D$, where $\log k_1(z)$ is the Poisson integral of $\frac{1}{2}\log\|P(\theta)^{-1}\|$ and $\log k_2(z)$ is the Poisson integral of $\frac{1}{2}\log\|P(\theta)\|$.

This result is an extension of the Wiener-Masani theorem (see [8]) which states:

If $P(\theta) = (p_{jk}(\theta))$ is a positive definite $n \times n$ matrix for each $\theta \in [0,2\pi)$ such that $p_{jk}(\theta)$ belongs to $L^1(0,2\pi)$, $j,k = 1,2,\ldots,n$, and

$$(5.11) \qquad -\infty < \int_0^{2\pi} \log\det P(\theta)\, d\theta,$$

then $P(\theta)$ can be factored as $P(\theta) = b(e^{i\theta})^* b(e^{i\theta})$, where $b(e^{i\theta}) = (b_{jk}(e^{i\theta}))$ is such that $b_{jk}(e^{i\theta})$ belongs to $L^2(0,2\pi)$, $j,k = 1,2,\ldots,n$, and each $b_{jk}(e^{i\theta})$ has a Fourier series of power series type.

That (5.8) implies this theorem is immediate: Since $\|P(\theta)\|$ is the largest proper value of $P(\theta)$ and $\|P(\theta)^{-1}\|$ is the reciprocal of the smallest proper value of $P(\theta)$, condition (5.11), together with the integrability of the $p_{jk}$'s, imply the integrability of $\log\|P(\theta)\|$ and $\log\|P(\theta)^{-1}\|$. Consequently, theorem (5.8) can be applied and we obtain the factorization (5.9). Since $b(z)$ is analytic, it has a Fourier series of power series type on the boundary. From (5.10) we obtain $\|b(e^{i\theta})\| \le \|P(\theta)\|^{1/2}$. Thus, the integrability assumption on the $p_{jk}$'s gives us the square integrability of the $b_{jk}$'s.

It is not hard to extend theorem (5.8) to the case where the boundary spaces are separable infinite dimensional Hilbert spaces. One then obtains the extension of the Wiener-Masani theorem that was established by Devinatz. This extension has been done by one of our students, S. Bloom, and will appear elsewhere.

Other applications and observations concerning these results can be found in [3].

The motivation for the choice of the material presented in these lectures at the University of Maryland was of a pedagogical nature. As we stated before, we considered the boundary spaces to be finite dimensional in order to avoid considerable technical difficulties of finding an appropriately large common subspace of these boundary spaces. Moreover, this choice also simplified all questions concerning "duality results." Another aspect of the theory we have not discussed involves interpolation of nonlinear operators on Banach spaces. In fact, it is the analyticity of $(T_zF)(z)$ (and not the linearity of $T_z$) that is of basic importance in our interpolation theorem. Examples of nonlinear analytic operators arise frequently in mathematics; for example, it can be shown that the functions arising in the Riemann mapping theorem vary analytically with the domain (if we have the "correct" parametrizations). Various regularity results of these functions can be proved by using our interpolation theory. It is our intention to publish, in the near future, a paper containing the general theory and more applications.

## REFERENCES


[1] Beckner, W. Inequalities in Fourier Analysis, Ann. of Math. 102 (1975), pp. 159-182.

[2] Calderón, A.P. Intermediate Spaces and Interpolation, the Complex Method, Studia Math. (1964), pp. 113-190.

[3] Coifman, R., Cwikel, M., Rochberg, R., Sagher, Y., and Weiss, G. Complex Interpolation for Families of Banach Spaces, Proceedings of Symposia in Pure Mathematics, vol. 35, Part 2, A.M.S. publication (1979), pp. 269-282.

[4] Dunford, N. and Schwartz, J.T. Linear Operators, Interscience Publishers, New York (1958).

[5] Hardy, G.H., Littlewood, J.E. and Pólya, G. Inequalities, Cambridge Univ. Press, London (1934).

[6] Stein, E.M. Interpolation of Linear Operators, Trans. Amer. Math. Soc., vol. 83, No. 2 (1956), pp. 482-492.

[7] Weissler, F.B. Hypercontractive Estimates for Semigroups, Proceedings of Symposia in Pure Math., vol. 35, Part 1, A.M.S. publication (1979), pp. 159-162.

[8] Wiener, N. and Akutowicz, E.J. A Factorization of Positive Hermitian Matrices, J. Math. and Mech. 8(1959), pp. 111-120.

[9] Wilansky, A. Functional Analysis, Blaisdell Publ. Co., New York (1964).

[10] Zygmund, A. Trigonometric Series, Cambridge Univ. Press, Cambridge (1959).

# MAXIMAL FUNCTIONS: A PROBLEM OF A. ZYGMUND

A. Córdoba
Princeton University

In 1910 H. Lebesgue extended the fundamental theorem of calculus in his well-known paper, Sur l'intégration des fonctions discontinues, (Ann. Ec. Norm. 27): Let $f$ be a locally integrable function on $\mathbb{R}^n$. Then

$$\lim_{r\to 0} \frac{1}{\mu[B(x;r)]} \int_{B(x;r)} f(y)d\mu(y) = f(x), \quad \text{a.e. } x,$$

where $\mu$ denotes Lebesgue measure in $\mathbb{R}^n$.

The quantitative interpretation of this result was obtained by Hardy and Littlewood in 1930 (A maximal theorem with function-theoretic applications, Acta Math. 54). Given a locally integrable function $f$ let us define

$$Mf(x) = \sup_{r>0} \frac{1}{\mu[B(x;r)]} \int_{B(x;r)} |f(y)|d\mu(y).$$

Then it follows that there exists a universal constant $C<\infty$ such that

$$\mu\{Mf(x) > \alpha\} \leq C \frac{\|f\|_1}{\alpha}.$$

Later on, E. Stein (Limits of sequences of operators, Annals. of Math. 1960) proved that, under very general conditions, the qualitative and the quantitative results mentioned above are in fact equivalent.

It is interesting to observe that if one replaces balls or cubes in the statement of the Lebesgue theorem by more general families of sets, for example parallelepipeds in $\mathbb{R}^n$ with sides parallel to the coordinate axes, then the differentiation theorem is false in general for integrable functions (Saks 1933). In 1935 Jessen, Marcinkiewicz and Zygmund showed that, in $\mathbb{R}^n$, we can differentiate the integral of $f$ with respect to the basis of intervals consisting of parallelepipeds with sides parallel to the coordinate axes, so long as $f$ belongs locally to the space $L(\log^+ L)^{n-1}(\mathbb{R}^n)$. This result is the best possible in the sense of Baire category.

The theory of differentiation of integrals has been closely related to the covering properties of families of sets. A classical example is the use of the Vitali covering lemma in the proof of the differentiation theorem of Lebesgue. In [1] a very precise interpretation of this relationship is given, and [3] contains a geometric proof of the result of Jessen, Marcinkiewicz and Zygmund by using a covering lemma of exponential type for intervals.

Given a positive function $\Phi$ on $\mathbb{R}^2$, monotonic in each variable separately, consider the differentiation basis $B_\Phi$ in $\mathbb{R}^3$ defined by the two parameter family of parallelepipeds whose sides are parallel to the rectangular coordinate axes and whose dimensions are given by $s\times t\times\phi(s,t)$, where $s$ and $t$ are positive real numbers. In general, the differentiation properties of $B_\Phi$ must be, at least, not worse than $B_3$, the basis of all parallelepipeds in $\mathbb{R}^3$ whose sides have the directions of the coordinate axes and, of course, not better than $B_2$. We will show that, in fact, $B_\phi$ behaves like $B_2$ from the differentiation point of view as well as for the estimates for the corresponding maximal function and covering properties. I believe that A. Zygmund was the first mathematician to pose this problem after his 1935 paper in collaboration with B. Jessen and J. Marcinkiewicz. This result and its extensions to higher dimensions are useful to understand the behavior of Poisson kernels associated with certain symmetric spaces.

Results

Theorem. (a) $B_\Phi$ differentiates integrals of functions which are locally in $L(1+\log^+ L)(\mathbb{R}^3)$, that is

$$\lim_{\substack{R\Rightarrow x\\ R\in B_\Phi}} \frac{1}{\mu\{R\}}\int_R f(y)d\mu(y) = f(x), \quad \text{a.e. } x$$

so long as $f$ is locally in $L(1+\log^+ L)(\mathbb{R}^3)$, where $\mu$ denotes Lebesgue measure in $\mathbb{R}^3$.

(b) The associated maximal function

$$M_\Phi f(x) = \sup_{\substack{x\in R\\ R\in B_\Phi}} \frac{1}{\mu\{R\}}\int_R |f(y)|d\mu(y)$$

satisfies the inequality

$$\mu\{M_\Phi f(x)>\alpha>0\} \leq C\int_{\mathbb{R}^3} \frac{|f(x)|}{\alpha}\left\{1+\log^+\frac{|f(x)|}{\alpha}\right\}d\mu(x)$$

for some universal constant $C<\infty$. The proof is based on the following geometric lemma.

Covering lemma. Let $B$ be a family of dyadic parallelepipeds in $\mathbb{R}^3$

satisfying the following monotonicity property: If $R_1, R_2 \in B$ and the horizontal dimensions of $R_1$ are both strictly smaller than the corresponding dimensions of $R_2$, then the vertical dimension of $R_1$ must be less than or equal to the vertical dimension of $R_2$.

It follows that the family $B$ has the exponential type covering property: Given $\{R_\alpha\} \subset B$ one can select a subfamily $\{R_j\} \subset \{R_\alpha\}$ such that

(i) $\mu\{\cup R_\alpha\} \le C\mu\{\cup R_j\}$, and

(ii) $\int_{\cup R_j} \exp(\Sigma \chi_{R_j}(x))d\mu(x) \le C\mu\{\cup R_j\}$

for some universal constant $C<\infty$.

<u>Application</u>. Consider

$$R^3 = \{X = \begin{pmatrix} x_1 & x_3 \\ x_3 & x_2 \end{pmatrix}, \text{ real, symmetric, } 2\times 2\text{-matrices}\},$$

and the cone $\Gamma = \{Y \in \mathbb{R}^3,$ positive definite$\}$. Then $T_\Gamma$ = tube over $\Gamma$ = Siegel's upper half-space = $\{X + iY,\ Y$ positive definite$\}$.

For each integrable function $f$ on $\mathbb{R}^3$ we have the "Poisson integral," $u(X + iY) = P_Y * f(X)$, $Y \in \Gamma$, where

$$P_Y(X) = C[\det Y]^{3/2}/|\det(X+iY)|^3$$

and we may ask the following question: For which functions $f$ is it true that $u(X + iY) \to f(X)$, a.e. $x$ when $Y \to 0$?.

It is a well-known fact that if $Y = y\cdot I = \begin{pmatrix} y & 0 \\ 0 & y \end{pmatrix} \to 0$, then $u(X + iY) \to f(X)$, a.e. $x$, for integrable functions $f$.

On the other hand if $Y \to 0$ without any restriction, then a.e. convergence fails for every class $L^P(\mathbb{R}^3)$, $1 \le P \le \infty$. Here we can settle the case

$$Y = \begin{pmatrix} y_1 & 0 \\ 0 & y_2 \end{pmatrix} \to 0$$

because an easy computation shows that $Mf(X) = \sup_Y |u(X + iY)|$ where

$$Y = \begin{pmatrix} y_1 & 0 \\ 0 & y_2 \end{pmatrix},$$

is majorized, in a suitable sense, by $M_\phi f(x)$ with $\phi(s,t) = (s\cdot t)^{1/2}$.

Therefore, we have a.e. convergence for the class $L(1+\log^+ L)(\mathbb{R}^3)$.

Proofs.

[A] Proof of the covering lemma.

We can assume that the given family satisfies the condition $\mu\{\cup R_\alpha\} < \infty$, otherwise there is nothing to be proved. Therefore we can also assume that $\{R_\alpha\}$ is finite and no one of its members is contained in the union of the others.

Let us choose $R_1$ to be an element of $\{R_\alpha\}$ with biggest vertical side. Assuming that we have chosen $R_1,\dots,R_{j-1}$, let $R_j$ be an element of $\{R_\alpha\}$ such that its vertical side is the biggest possible among the $\alpha$'s that satisfy

$$\frac{1}{\mu\{R_\alpha\}}\int_{R_\alpha}\exp\Big(\sum_{k=1}^{j-1}\chi_{R_k}(x)\Big)dx \;\le\; 1+e^{-1}.$$

The subfamily, $\{R_j\}_{j=1,\dots,M}$ obtained in this way, satisfies

$$\int_{\bigcup_{j=1}^{M}R_j}\exp\Big(\sum_{j=1}^{M}\chi_{R_j}(x)\Big)dx \;=\; \int_{\bigcup_{j=1}^{M-1}R_j-R_M}\exp\Big(\sum_{j=1}^{M-1}\chi_{R_j}(x)\Big)dx + e\int_{R_M}\exp\Big(\sum_{j=1}^{M-1}\chi_{R_j}(x)\Big)dx \le$$

$$\le \int_{\bigcup_{j=1}^{M-1}R_j}\exp\Big(\sum_{j=1}^{M-1}\chi_{R_j}(x)\Big)dx + (1+e)\mu\{R_M\} \;\le$$

$$\le\; (1+e)\Sigma\mu(R_j) \;\le\; \frac{(1+e)(e-1)}{e-1-e^{-1}}\,\mu(\cup R_j).$$

Next, given $R\in\{R_\alpha\}-\{R_j\}$, we have

$$\frac{1}{\mu\{R\}}\int_R \exp(\Sigma'\chi_{R_j}(x))dx \;\ge\; 1+e^{-1},$$

where $\Sigma'$ is extended over all parallelepipeds $R_j$ with bigger vertical dimension than $R$.

Let us rearrange the rectangles appearing in $\Sigma'$ in the following way

$$\Sigma'\chi_{R_j} \;=\; \chi_{R_1}+\dots+\chi_{R_p}+\chi_{R_{p+1}}+\dots+\chi_{R_q}$$

where the $R_j$'s, $j = 1,\dots,p$ have s-dimension ≥ s-dimension of $R$ and those $R_j$'s with $j = p+1,\dots,q$, have t-dimension ≥ t-dimension of $R$.

Then

$$1 + e^{-1} \le \frac{1}{\mu\{R\}} \int_R \exp(\Sigma' \chi_{R_j}) dx$$

$$= \sum_{r,s=0}^{\infty} \frac{1}{r!} \frac{1}{s!} \sum_{k_1,\dots,k_r=1}^{p} \sum_{\ell_1,\dots,\ell_s=p+1}^{q} \frac{1}{\mu\{R\}} \int_R \chi_{R_{k_1}} \cdots \chi_{\ell_s} dx$$

$$= \sum_{r,s=0}^{\infty} \frac{1}{r!} \frac{1}{s!} \sum_{k_1,\dots,k_r=1}^{p} \sum_{\ell_1,\dots,\ell_s=p+1}^{q} \frac{\mu\{R_{k_1} \cap \dots \cap R_{k_r} \cap R_{\ell_1} \cap \dots \cap R_{\ell_s} \cap R\}}{\mu\{R\}} .$$

If $\mu\{R_{k_1} \cap \dots \cap R_{k_r} \cap R_{\ell_1} \cap \dots \cap R_{\ell_s} \cap R\} \neq 0$ then the intersection must be of the form shown in the figure, where "the block corresponding to $\varepsilon$" = the intersection of the $R_j$'s whose t-dimension is bigger than the t-dimension of $R$, and analogously for the block $\delta$.

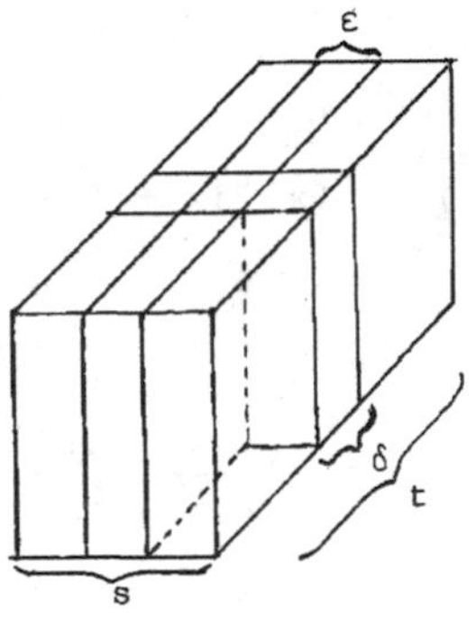


Therefore,

$$\frac{\mu\{R_{k_1} \cap \dots \cap R_{\ell_s} \cap R\}}{\mu\{R\}} = \frac{\varepsilon \times \delta}{s \times t} .$$

Given $P = (x_0, y_0, z_0) \in R$, consider

$$I_P^1 = \{(x, y_0, z_0) \in R\},$$

$$I_P^2 = \{(x_0, y, z_0) \in R\}.$$

It happens, again by the monotonicity, that

$$\frac{|R_{k_1}\cap\ldots\cap R_{k_r}\cap I_p^{\,1}|}{|I_p^{\,1}|} = \frac{\varepsilon}{s}, \quad \text{and}$$

$$\frac{|R_{\ell_1}\cap\ldots\cap R_{\ell_s}\cap I_p^{\,2}|}{|I_p^{\,2}|} = \frac{\delta}{t}.$$

Therefore,

$$1 + e^{-1} \le \frac{1}{\mu(R)}\int_R \exp(\Sigma'\chi_{R_j})$$

$$\le \sum_{r,s=0}^{\infty} \frac{1}{r!}\frac{1}{s!} \sum_{k_1\ldots k_r=1}^{p} \sum_{\ell_1\ldots\ell_s=p+1}^{q} \frac{|R_{k_1}\cap\ldots\cap R_{k_r}\cap I_p^{\,1}|}{|I_p^{\,1}|}$$

$$\cdot \frac{R_{\ell_1}\cap\ldots\cap R_{\ell_s}\cap I_p^{\,2}}{|I_p^{\,2}|}$$

$$\le \left[\frac{1}{|I_p^{\,1}|}\int_{I_p^{\,1}} \exp(\Sigma\chi_{R_j})\right] \cdot \left[\frac{1}{|I_p^{\,2}|}\int_{I_p^{\,2}} \exp(\Sigma\chi_{R_j})\right].$$

Here, $|\ |$ denotes 1-dimensional Lebesgue measure. Therefore if $M_x$, $M_y$ denote the extensions to $\mathbb{R}^3$ of the one-dimensional Hardy-Littlewood maximal function in the $x$ and $y$ directions respectively, we have:

$$R \subset \{M_x(\exp(\Sigma\chi_{R_j})) \cdot M_y(\exp(\Sigma\chi_{R_j})) \ge 1 + e^{-1}\}$$

$$\subset \{M_x(\exp(\Sigma\chi_{R_j}) \ge \sqrt{1 + e^{-1}}\} \cup \{M_y(\exp(\Sigma\chi_{R_j})) \ge \sqrt{1 + e^{-1}}\}.$$

Thus,

$$\mu(\cup R_\alpha) \le \frac{2}{\sqrt{1+e^{-1}}}\int_{\cup R_j} \exp(\Sigma\chi_{R_j}) \le C\mu(\cup R_j).$$

Q.E.D.

[B] Proof of the theorem.

It is a well-known fact that statements (a) and (b) are equivalent, and the implication (b) ⇒ (a) is the easier of the two. The strategy of the proof is as follows: First we observe that if the family B satisfies the hypothesis of the covering lemma and if M denotes the maximal function associated with B, then M satisfies (b); next we find such a family B with the property that $M_\Phi f \le CMf$ for some universal constant C.

Let B satisfy the hypotheses of the covering lemma, and define

$$Mf(x) = \sup_{\substack{x\in R^* \\ R\in B}} \frac{1}{\mu(R)} \int_R |f(y)|d\mu(y),$$

where $R^*$ is the result of dilating R by a factor of 3 with respect to its center.

Claim. There exists a constant $C<\infty$ such that

$$\mu\{Mf(x) > \lambda\} \le C \int_{\mathbb{R}^3} \frac{|f(x)|}{\lambda} \left\{1 + \overset{+}{\log} \frac{|f(x)|}{\lambda}\right\} d\mu(x).$$

Proof of the claim.

The set $\{Mf(x) > \lambda\}$ is a union of parallelepipeds $\{R_\alpha^*\}$ such that

$$\frac{1}{\mu\{R_\alpha\}} \int_R |f(y)|d\mu(y) \ge \lambda.$$

We can apply the covering lemma to show the existence of the subfamily $\{R_j\}$ with the prescribed properties. We have

$$\mu\{UR_j^*\} \le 27\ \Sigma\mu(R_j) \le 27 \int_{UR_j} \frac{|f(x)|}{\lambda} \Sigma\chi_{R_j}(x)\, d\mu(x).$$

Next, observe that if u and v are positive real numbers, it follows that $u \cdot v \le u \log u + \exp(v-1)$. Furthermore, for every $\varepsilon > 0$ there exists a constant $C_\varepsilon < \infty$ such that $u \cdot v \le C_\varepsilon u(1 + \overset{+}{\log}\, u) + \exp(\varepsilon v-1)$. In particular,

$$\mu\{UR_j^*\} \le C_\varepsilon \int_{UR_j} \frac{|f(x)|}{\lambda} \{1 + \overset{+}{\log} \frac{|f(x)|}{\lambda}\}d\mu(x) + e^{-1} \int_{UR_j} \exp(\varepsilon\Sigma\chi_{R_j}(x))d\mu(x).$$

But

$$\int_{UR_j} \exp(\varepsilon\Sigma\chi_{R_j}(x))d\mu(x) \le [\mu(UR_j)]^{1-\varepsilon}\left(\int_{UR_j} \exp(\Sigma\chi_{R_j}(x))\right)^{\varepsilon} d\mu(x) \le C^{\varepsilon}\mu(UR_j),$$

where $C$ is the constant appearing in the covering lemma. To finish, we just choose $\varepsilon > 0$ so that $C^{\varepsilon}e^{-1} \le \frac{1}{2}$.

Finally we must show how to get the family $B$ from $B_{\phi}$. First of all we can reduce the values of $s$ and $t$ so that they are in the set $\{2^{+n}\}_{n\in Z}$; then we define $\tilde{\phi}(2^k,2^{\ell}) = 2^m$ if $2^{m-1} < \phi(2^k,2^{\ell}) \le 2^m$. It is clear that if $\tilde{B}$ is the family of parallelepipeds obtained in this way then each element $R$ of $B_{\Phi}$ is contained in an element $\tilde{R}$ of $\tilde{B}$ in such a way that $\mu\{\tilde{R}\} \le 8\mu\{R\}$. Next, we consider in the horizontal plane $z = 0$, the family of dyadic rectangles, and to each one of these rectangles we attach a parallelepiped, having the rectangle as horizontal base and with vertical dimension given by $\tilde{\Phi}$. Then we translate, in the vertical direction, each one of these parallelepipeds by integer multiples of its vertical length. The family obtained is $B$. Clearly $M_{\phi}f(x) \le 8Mf(x)$.

Q.E.D.

## References


[1] A. Córdoba, On the Vitali covering properties of a differentiation basis, Studia Math. 57 (1976), 91-95.

[2] ——————, s×t×ϕ(s,t), Mittag-Leffler Institute report 9, 1978.

[3] ——————, and R. Fefferman, A geometric proof of the strong maximal theorem, Annals of Math. 102, 1975.

[4] B. Jessen, J. Marcinkiewicz and A. Zygmund, Note on the differentiation of multiple integrals, Fund. Math. 25, 1935.

# MULTIPLIERS OF $F(L^p)$

A. Córdoba
Princeton University

## I. The disc multiplier and related problems in $\mathbb{R}^2$

I would like to present today three results in Fourier Analysis which are closely related to the spherical Cesaro means of multiple Fourier Series. Some of these results can be extended to higher dimensions but I will restrict myself to $\mathbb{R}^2$ where it is possible now to present a more complete description.

Consider the family of Fourier multipliers defined by the formula

$$\widehat{T_\lambda f}(\xi) = (1-|\xi|^2)_+^\lambda \hat{f}(\xi), \quad \lambda \geq 0, \quad f \in S(\mathbb{R}^2).$$

If $\lambda > \frac{1}{2}$ then $T_\lambda f = K_\lambda * f$ where $K_\lambda$ is an integrable kernel, so the $L^p$-theory is very easy in this case, and we shall concentrate only on what happens where $\lambda \leq \frac{1}{2}$ and $K_\lambda$ fails to be in $L^1$, ($T_0$ is the multiplier associated to the unit disc). Now C. Herz [7] observed that $T_\lambda$ is unbounded outside the range

$$p(\lambda) = \frac{2n}{n+1+2\lambda} < p < \frac{2n}{n-1-2\lambda} = p'(\lambda).$$

The known theory of $T_\lambda$ in $\mathbb{R}^2$ can be summarized in the following theorems.

Theorem (A)

(a) $T_0$ is only bounded on $L^2(\mathbb{R}^2)$, (C. Fefferman [5]).

(b) If $\frac{1}{2} \geq \lambda > 0$, then $T_\lambda$ is bounded on $L^p(\mathbb{R}^2)$, $p(\lambda) < p < p'(\lambda)$, (L. Carleson and P. Sjölin [1]).

Given $N \geq 1$, consider $B_N$ = {rectangles of eccentricity $\leq N$} (arbitrary direction) and the associated maximal function

$$Mf(x) = \sup_{x \in R \in B_N} \frac{1}{\mu(R)} \int_R |f(y)|dy.$$

Theorem (B). There exists a constant C, independent of N, such that

$$\mu\{Mf(x) > \alpha > 0\} \leq C(\log 3N) \frac{\|f\|_2^2}{\alpha^2},$$

for every $f \in L^2(\mathbb{R}^2)$.

The third result is a restriction theorem for the Fourier transform.

Theorem (C). Let $f \in L^p(\mathbb{R}^2)$, $1 \le p < \frac{4}{3}$. Then the Fourier transform $\hat{f}$ restricts to an $L^q$ function on the unit circle $S^1$, where $\frac{1}{q} \ge 3[1 - \frac{1}{p}]$, and satisfies the a priori inequality

$$\|\hat{f}\|_{L^q(S^1)} \le C_{p,q}\|f\|_{L^p(\mathbb{R}^2)}$$

(C. Fefferman and E. Stein [5], A. Zygmund [8]).

Strategy

The multiplier $m_\lambda(\xi) = (1-|\xi|^2)_+^\lambda$ seems very complicated and one of our first tasks is to find out which are the basic blocks of the Calderón-Zygmund theory corresponding to $m_\lambda$. Since $m_\lambda$ is radial and basically constant on thin annuli it seems reasonable to decompose

$$m_\lambda(\xi) = \sum_0^\infty (1-|\xi|^2)_+^\lambda \varphi_k(|\xi|)$$

where $\varphi_k$, $k \ge 1$, is a smooth function supported in the interval $[1 - 2^{-k}, 1 - 2^{-k-2}]$ such that $|D^\alpha \varphi_k| \le C_\alpha 2^{k\alpha}$, where $C_\alpha$ is independent of $k$, and

$$\sum_{k=1}^\infty \varphi_k \equiv 1 \text{ on } [\tfrac{1}{2}, 1], \quad \varphi_0 = 1 - \sum_{k=1}^\infty \varphi_k.$$

Then

$$(1-|\xi|^2)_+^\lambda \varphi_k(|\xi|) \simeq 2^{-k\lambda}\varphi_k(|\xi|)$$

and the problem is reduced to getting good estimates for the growth, as $k \to \infty$, of the norm of the multipliers associated with the function $\varphi_k(|\xi|)$. For example, the Carleson-Sjölin result will follow very easily if one can show that the operator

$$\widehat{T^k f}(\xi) = \varphi_k(|\xi|)\hat{f}(\xi)$$

satisfies the inequality

$$\|T^k f\|_4 \le Ck^{\frac{1}{4}}\|f\|_4, \quad \forall f \in S(\mathbb{R}^2)$$

because interpolation with the obvious estimate

$$\|T^k f\|_\infty \le C2^{\frac{k}{2}}\|f\|_\infty$$

yields

$$\|T^k f\|_p \leq Ck^{\frac{t}{4}} 2^{\frac{k(1-t)}{2}} \|f\|_p.$$

If $\frac{1}{p} = \frac{t}{4}$ and if $4 < p < \frac{4}{1-2\lambda}$, then $1 - t < 2\lambda$, and therefore the series

$$\sum_{k=1}^{\infty} 2^{-k\lambda} k^{\frac{1}{4}} 2^{\frac{k(1-t)}{2}}$$

converges, proving the Carleson-Sjölin theorem.

Statement and Proofs of Some Results

(a) Suppose that $\varphi : \mathbb{R} \longrightarrow \mathbb{R}$ is a smooth function supported in $[-1, +1]$, and consider the family of Fourier multipliers $S_\delta$, where $\delta > 0$ is small, defined by the formula

$$\widehat{S_\delta f}(\xi) = \varphi(\delta^{-1}(|\xi|-1)) \cdot \hat{f}(\xi),$$

for rapidly decreasing smooth functions $f$.

Theorem 1. There exists a constant $C$, independent of $\delta$, such that

$$\|S_\delta f\|_4 \leq C(\log(1/\delta))^{\frac{1}{4}} \|f\|_4$$

for every $f \in S(\mathbb{R}^2)$.

(b) Given real numbers $N \geq 1$ and $a > 0$, consider the family $B$ of rectangles in $\mathbb{R}^2$ with dimensions $a$ and $Na$, but with arbitrary direction. For a locally integrable function $f$ let us define the maximal function

$$Mf(x) = \sup_{x \in R \in B} \frac{1}{\mu(R)} \int_R |f(x)| d\mu(x)$$

where $\mu$ denotes Lebesgue measure in the plane.

Theorem 2. There exists a constant $C$, independent of $a$ and $N$, such that

$$\|Mf\|_2 \leq C(\log 2N)^{\frac{1}{2}} \|f\|_2, \quad \text{for every } f \in L^2(\mathbb{R}^2).$$

(c) Given two positive real numbers $N_1$ and $N_2$ let us consider the family of intervals in the y-axis, $\{I_j\}_{-\infty<j<+\infty}$ whose length is equal to $N_1$ and such that the distance between two consecutive intervals is equal to $N_2$. Denote by $E_j$ the horizontal strip in $\mathbb{R}^2$ whose projection onto the y-axis is the interval $I_j$. Given $f \in S(\mathbb{R}^2)$, we may define the g-function

$$g(f)(x) = (\sum_j |P_j f(x)|^2)^{\frac{1}{2}}, \quad \text{where } \widehat{P_j f}(\xi) = \chi_{E_j}(\xi)\hat{f}(\xi).$$

Theorem 3. For every $p \geq 2$ there exists a constant $C_p$ such that

$$\|g(f)\|_p \leq C_p \|f\|_p, \quad \forall f \in S(\mathbb{R}^2).$$

(d) Corollary. The operator $T_\lambda$, $\frac{1}{2} \geq \lambda > 0$, defined by

$$\widehat{T_\lambda f}(\xi) = (1-|\xi|^2)_+^\lambda \hat{f}(\xi)$$

for rapidly decreasing and smooth functions $f$, has a bounded extension to $L^p(\mathbb{R}^2)$ if and only if

$$\frac{4}{3+2\lambda} < p < \frac{4}{1-2\lambda}.$$

Proofs.

(a) Using smooth cut off functions we may decompose

$$\varphi = \sum_{j=0}^{3} \varphi_j,$$

where

$$\mathrm{supp}(\varphi_j) \subset \{z: -\frac{\pi}{3} + \frac{j\pi}{2} \leq \arg(z) < \frac{\pi}{3} + \frac{j\pi}{2}\}.$$

Since we may consider $\varphi_j$, $j = 1, 2, 3$, as a rotation of $\varphi_0$, and since the Fourier transform commutes with rotations, it is enough to prove that the multiplier $\varphi_0$ satisfies the estimate of Theorem 1.

Next, following [2], we consider a smooth partition of unity, $\{\Phi_j\}_{j=1,\dots,[\delta^{-1/2}]}$ in the unit circle, such that:

$$\Phi_j(\theta) = \Phi(\delta^{-\frac{1}{2}}(\theta - \frac{2\pi j}{[\delta^{-1/2}]})),$$

where $\Phi$ is a smooth function supported on the unit interval, with bounds independent of $\delta$. Therefore

$$\varphi_0(\xi) = \varphi_0(|\xi|e^{i\theta}) = \sum_j \varphi_0(\xi)\Phi_j(\theta) = \sum_j m_j(\xi).$$

If we define $\widehat{T_j f}(\xi) = m_j(\xi)\hat{f}(\xi)$, then we have

$$\int_{\mathbb{R}^2} |\sum_j T_j f(x)|^4 dx = \int_{\mathbb{R}^2} |\sum_{j,k} T_j f(x) T_k f(x)|^2 dx$$

$$= \int_{\mathbb{R}^2} |\sum_{j,k} \widehat{T_j f} * \widehat{T_k f}(\xi)|^2 d\xi \leq C \sum_{j,k} \int_{\mathbb{R}^2} |\widehat{T_j f} * \widehat{T_k f}(\xi)|^2 d\xi .$$

This last inequality holds because no point belongs to more than 16 sets of the family $\mathrm{Supp}(\widehat{T_j f}) + \mathrm{Supp}(\widehat{T_k f})$. Therefore,

$$\|\sum_j T_j f\|_4 \leq C\|(\sum_j |T_j f(x)|^2)^{\frac{1}{2}}\|_4$$

where $C$ is independent of $\delta$.

Next, we split the sum

$$(\sum_j |T_j f(x)|^2)^{\frac{1}{2}} \leq (\sum_j |T_{2j} f(x)|^2)^{\frac{1}{2}} + (\sum_j |T_{2j+1} f(x)|^2)^{\frac{1}{2}},$$

and we estimate each of the two sums separately.

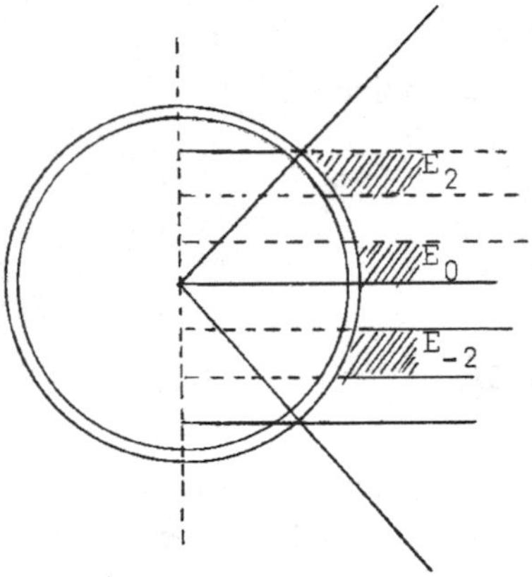


Figure 1

Since the supports of the multipliers $m_j$ are basically perpendicular to the direction of the $x$-axis, we may find a configuration, $\{E_{2j}\}$, horizontal strips, as in Theorem 3, with the property that

$$\chi_{E_{2j}}(\xi) \cdot m_{2j}(\xi) = m_{2j}(\xi).$$

Therefore,

$$(\sum |T_{2j} f(x)|^2)^{\frac{1}{2}} = (\sum |T_{2j} P_{2j} f(x)|^2)^{\frac{1}{2}}$$

where

$$\widehat{P_{2j}h}(\xi) = \chi_{E_{2j}}(\xi)\hat{h}(\xi).$$

An elementary computation (integration by parts) shows that the kernel $K_0$ of $T_0$ satisfies the following: For every pair of integers $p,q \geq 0$ there exists a finite constant $C_{p,q}$, independent of $\delta$, such that

$$|K_0(x,y)| \leq C_{p,q}\delta^{\frac{3}{2}}|\delta x|^{-p}|\delta^{\frac{1}{2}}y|^{-q}.$$

Therefore the operator $T_0$ is majorized by the positive operator whose kernel is given by

$$C\left\{\sum_{n=0}^{\infty} 2^{-n} \frac{1}{\mu(R_n^0)} \chi_{R_n^0}(x,y)\right\},$$

where $R_n^0 = \{(x,y): |x| \leq 2^n\delta^{-1}, |y| \leq 2^n\delta^{-1/2}\}$ for a suitable constant $C$, independent of $\delta > 0$. By an appropriate rotation we may get an analogous majorization for every operator $T_j$.

Due to the exponential decay factor $2^{-n}$ it is enough to show that, for each $n$, the $L^4$-norm of the function

$$\left(\sum_j \left|\frac{1}{\mu(R_n^{2j})} \chi_{R_n^{2j}}*P_{2j}f(x)\right|^2\right)^{\frac{1}{2}}$$

is dominated by $C(\log\delta)^{\frac{1}{4}}\|f\|_4$, for every $f \in L^4(\mathbb{R}^2)$. Given $w \geq 0$ in $L^2(\mathbb{R}^2)$ we have

$$\sum_j \int_{\mathbb{R}^2} \left|\frac{1}{\mu(R_n^{2j})} \chi_{R_n^{2j}}*P_{2j}f(x)\right|^2 w(x)dx$$

$$\leq \sum_j \int_{\mathbb{R}^2} |P_{2j}f(y)|^2 \frac{1}{\mu(R_n^{2j})} \chi_{R_n^{2j}}*w(y)dy$$

$$\leq \sum_j \int_{\mathbb{R}^2} |P_{2j}f(y)|^2 M_n w(y)dy \leq \|(\sum_j |P_{2j}f|^2)^{\frac{1}{2}}\|_4^2 \cdot \|M_n w\|_2,$$

where $M_n$ is the maximal function of Theorem 2, with

$$N = \delta^{-\frac{1}{2}} \text{ and } a = 2^n.$$

Therefore

$$\|(\Sigma|T_{2^j}f|^2)^{\frac{1}{2}}\|_4^2 \le \sup_{\|w\|_2\le 1}\int \sum_j |T_{2^j}P_{2^j}f(x)|^2 w(x)dx$$

$$\le C\sum_n 2^{-n}(\log 1/\delta)^{\frac{1}{2}}\|(\Sigma|P_{2^j}f|^2)^{\frac{1}{2}}\|_4^2 \le \tilde{C}(\log(1/\delta))^{\frac{1}{4}}\|f\|_4^2,$$

by Theorems 2 and 3.

An analogous argument works for the odd sum. Q.E.D.

This Theorem, with the bigger power $(\log 1/\delta)^{5/4}$, was proved in [2] for the first time. The proof presented here, using the g-function of Theorem 3, yields the best exponent, 1/4.

(b) Theorem 2 was proved in [2]. The exponent 1/2 cannot be improved, as the case $a = 1$, $f(x) = (1+|x|)^{-1}$, if $|x| \le N$ and $f(x) = 0$, if $|x| > N$ shows.

Proof of Theorem 2.

First of all, it is enough to prove the estimate for rectangles whose directions lie in the interval $[0,\pi/4]$. By using a convenient dilation we may also assume that $a = 1$. We divide the plane by vertical and horizontal lines, into a grid of squares of side N. The operator M acts "independently" on the squares of the grid, and therefore we can simplify the problem by considering only functions f supported on one of the squares of the grid. So, let Q be a square with sides parallel to the coordinate axes, and length = N, and suppose that $f \in L^2(Q)$. Then

$$Mf(x) = 0 \text{ if } x \in Q^*.^{(+)}$$

We decompose the square $Q^*$ into $9N^2$ small squares $\{Q_{ip}\}$, of side = 1, by vertical and horizontal lines. The point is that for every square $Q_{ip}$, one can find a rectangle $R_{ip}$ (with direction in the interval $[0,\pi/4]$, and dimensions $1 \times N$) such that

$$Q_{ip} \cap R_{ip} \ne \emptyset \text{ and } Mf(x) \le 2\frac{1}{|R_{ip}|}\int_{R_{ip}}|f(y)|dy \cdot \chi_{Q_{ip}}(x).$$

Therefore, if for a fixed f we define the linear operator:

---

(+) $Q^*$ denotes the square expanded by the factor 3.

$$T_f(g)(x) = \sum_{i,p} \frac{1}{|R_{ip}|} \int_{R_{ip}} g(y)dy \cdot \chi_{Q_{ip}}(x),$$

we have $Mf(x) \leq 2T_f(|f|)(x)$. In order to prove the theorem, it is enough to prove that

$$\|T_f(g)\|_2 \leq C(\log 3N)^{\frac{1}{2}}\|g\|_2 \forall g \in L^2(Q^*),$$

where $C$ is independent of $f$ and $N$.

Thus we have linearized the problem, and we can consider the adjoint $T_f^*$ of $T_f$, which is given by:

$$T_f^*(h)(x) = \sum_{i,p} \frac{1}{R_{ip}} (\int_{Q_{ip}} h(y)dy)\chi_{R_{ip}}(x).$$

Now, given $h \in L^2(Q^*)$ we have the decomposition $h = h_1 + \dots + h_{3N}$, where $h_i = h|_{E_i}$ is the restriction of $h$ to the vertical strip $E_i$ of width 1. Then in order to prove that

$$\|T_f^*(h)\|_2 \leq C(\log 3N)^{\frac{1}{2}}\|h\|_2$$

it is enough to show that

$$\|T_f^*(h_i)\|_2 \leq CN^{-\frac{1}{2}}(\log 3N)^{\frac{1}{2}}\|h_i\|_2, \quad i = 1,\dots,3N,$$

since this implies that

$$\|T_f^*(h)\|_2 \leq \sum_i \|T_f^*(h_i)\|_2 \leq CN^{-\frac{1}{2}}(\log 3N)^{\frac{1}{2}}\sum_i \|h_i\|_2 \leq C(\log 3N)^{\frac{1}{2}}\|h\|_2.$$

Suppose that the function $h$ is defined on the strip $E_i$. We decompose $E_i$ into $3N$ squares $\{Q_{ip}\}_{p=1,\dots,3N}$, each of side 1, and also we decompose the function $h = \Sigma h_p$ where $h_p = h|Q_{ip}$. We have

$$T_f^*(h)(x) = \sum_p T_f^*(h_p)(x) = \sum_p \frac{1}{|R_{ip}|} \int_{Q_{ip}} h_p(y)dy\chi_{R_{ip}}(x),$$

which implies that

$$|T_f^*(h)(x)| \leq \frac{1}{N} \sum_{p=1}^{3N} \|h_p\|_2 \chi_{R_{ip}}(x).$$

Therefore,

$$\int |T_f^*(h)(x)|^2 dx \leq \frac{1}{N^2} \sum_{p,q} \|h_p\|_2 \|h_q\|_2 |R_{ip} \cap R_{iq}|,$$

and an easy computation shows that $|R_{ip} \cap R_{iq}| \leq C \frac{N}{1+|p-q|}$ .

This implies that

$$\|T_f^*(h)\|_2^2 \leq C\frac{1}{N} \sum_{p,q=1}^{3N} \frac{\|h_p\|_2 \|h_q\|_2}{1+|p-q|} \leq CN^{-1} \log 3N \cdot \|h\|_2^2 \quad \text{Q.E.D.}$$

(c) We have used a two dimensional version of Theorem 3 to prove Theorem 1 although the latter theorem is basically one-dimensional. In what follows we sketch the proof of Theorem 3, assuming without loss of generality, that $N_1 = 2$, $N_2 = N$ and the first interval $I_0$ is centered at the origin: $I_j = (w_j - 1, w_j + 1)$, where $w_j = j(N+1)$.

Let $\psi$ be a smooth function such that $\psi \equiv 1$ on $I_0$ and $\psi \equiv 0$ outside the interval

$$\left(-1 - \frac{N}{4}, \ 1 + \frac{N}{4}\right)$$

and let $\psi_j(t) = \psi(t - w_j)$ and $\widehat{S_j f}(\xi) = \psi_j(\xi)\hat{f}(\xi)$.

Lemma.

$$\|(\sum_j |S_j f(x)|^2)^{\frac{1}{2}}\|_p \leq C_p \|f\|_p,$$

for every $p \geq 2$.

Proof. For each $\theta$, $0 \leq \theta \leq 2\pi$, consider the multiplier

$$\widehat{T_\theta f}(\xi) = \sum_j e^{i\theta j} \psi_j(\xi)\hat{f}(\xi)$$

and observe that its kernel $\mu_\theta$ is a measure of finite total variation, uniformly in $\theta$. Therefore, for every $\theta$ we have

$$\left(\int |T_\theta f(x)|^p dx\right)^{\frac{1}{p}} \leq C_p \|f\|_p.$$

Integrate with respect to $\theta$, and observe that if $p \geq 2$, then

$$C_p^p \int |f(x)|^p dx \geq \frac{1}{2\pi} \int_0^{2\pi} \left(\int |T_\theta f(x)|^p dx\right) d\theta$$

$$= \int \frac{1}{2\pi} \int_0^{2\pi} |T_\theta f(x)|^p d\theta dx$$

$$\geq \int \left(\frac{1}{2\pi} \int_0^{2\pi} |T_\theta f(x)|^2 d\theta\right)^{\frac{p}{2}} dx = \int \left(\sum_j |S_j f(x)|^2\right)^{\frac{p}{2}} dx.$$

Q.E.D.

The proof of Theorem 3 is now easy to obtain.

(d) Let us discuss Theorem C, whose proof follows the same pattern as that of Theorem A.

Theorem. If $f \in L^p(\mathbb{R}^2)$, $1 \leq p < \frac{4}{3}$, then $\hat{f}$ exists almost everywhere on $|\xi| = 1$, and we have

$$\left(\int_{|\xi|=1} |\hat{f}(\xi)|^q d\sigma(\xi)\right)^{\frac{1}{q}} \leq A_{p,q} \|f\|_p$$

where $q = \frac{1}{3}p' = \frac{1}{3}\frac{p}{p-1}$.

Proof. It is enough to show that

$$\lim_{\delta \to 0} \left(\delta^{-1} \int_{1-\frac{\delta}{2} \leq |\xi| \leq 1+\frac{\delta}{2}} |\hat{f}(\xi)|^q d\xi\right)^{\frac{1}{q}} \leq A_{p,q} \|f\|_p$$

under the same conditions as those of the theorem.

By duality it is enough to prove the following estimate:

$$\delta^{-1} \|\Sigma a_j \hat{\varphi}_j\|_{p'} \leq C(\Sigma |a_j|^{q'} \delta^{\frac{1}{2}})^{\frac{1}{q'}}$$

where $p'$, $q'$ are conjugate exponents to $p$ and $q$ respectively, the $a_j$'s are complex numbers and the $\varphi_j$'s are characteristic functions of the "blocks" ($\delta^{1/2}$-in the tangential direction, $\delta$-in the normal direction) of the canonical partition of the annulus $1-\delta/2 \leq |\xi| \leq 1+\delta/2$. Observe that $p' > 4$. Therefore,

$$\left(\int \Big|\sum_{j,k=1}^{\delta^{-\frac{1}{2}}} a_j a_k \hat{\varphi}_j \hat{\varphi}_k\Big|^{\frac{p'}{2}} dx\right)^{\frac{2}{p'} \cdot \frac{1}{2}} \leq \left(\int \Big|\sum_{j,k=1}^{\delta^{-\frac{1}{2}}} a_j a_k \varphi_j * \varphi_k\Big|^r dx\right)^{\frac{1}{r} \cdot \frac{1}{2}}.$$

Where $\frac{1}{r} + \frac{2}{p'} = 1$. Since the supports of $\{\varphi_j * \varphi_k\}$ are "almost disjoint," and $\mu\{\mathrm{supp}(\varphi_j * \varphi_k)\} \leq \delta^{3/2}(|j-k|+1)$, it follows that

$$\|\varphi_j * \varphi_k\|_\infty \le \frac{\delta^{\frac{3}{2}}}{|j-k|+1}.$$

Hence we have

$$\|\Sigma a_j \hat{\varphi}_j\|_{p'} \le C\delta^{\frac{3}{4}(1+\frac{1}{r})} \left( \sum_{j,k=1}^{\delta^{-\frac{1}{2}}} \frac{|a_j|^r |a_k|^r}{(1+|j-k|)^{r-1}} \right)^{\frac{1}{2} \cdot \frac{1}{r}}$$

A fractional integration argument yields

$$\|\Sigma a_j \hat{\varphi}_j\|_{p'} \le C\delta^{\frac{3}{4}(1+\frac{1}{r})} \left[ \left( \Sigma |a_j|^{\frac{2r}{3-r}} \right)^{3-r} \right]^{\frac{1}{2r}} \le C\delta (\Sigma |a_j|^{q'} \delta^{\frac{1}{2}})^{\frac{1}{q'}}. \quad \text{Q.E.D.}$$

Observe that the only property used of the unit circle is its positive curvature. Therefore, we have in fact, proved a more general theorem.


References

[1] Carleson, L. and Sjölin, P., Oscillatory integrals and a multiplier problem for the disc, Studia Math. 1972.

[2] Córdoba, A., The Kakeya maximal function and the spherical summation multipliers, Amer. J. of Math. 1977.

[3] ——————, A note on Bochner-Riesz operators, Duke. Math. J. 1979.

[4] Fefferman, C., A note on spherical summation multipliers, Isr. J. of Math. 1973.

[5] ——————, The multiplier problem for the ball, Annals of Math. 1972.

[6] ——————, Inequalities for strongly singular convolution operators, Acta. Math. 1970.

[7] Herz, C., On the mean inversion of Fourier and Hankel transforms, P.N.A.S. 1954.

[8] Zygmund, A., On Fourier coefficients and transforms of functions of two variables, Studia Math. 1974.


## II. Further Results.

Let us begin by considering a sharper version of the Carleson-Sjölin theorem. Let $\chi_p$ denote the characteristic function of a regular polygon of N sides in $\mathbb{R}^2$ and consider the Fourier multiplier defined by $\widehat{Tf}(\xi) = \chi_P(\xi)\hat{f}(\xi)$.

Theorem 1. For each $p$ such that $\frac{4}{3} \leq p \leq 4$, there exist constants $C_p$ and $a(p)$, independent of $N$, such that

$$\|Tf\|_p \leq C_p(\log N)^{a(p)}\|f\|_p$$

for every $f \in S(\mathbb{R}^2)$.

This theorem is a consequence of the following maximal function result: Define

$$Mf(x) = \sup_{x \in R} \frac{1}{|R|}\int_R |f(y)|dy$$

where the "Sup" is taken over all rectangles in $\mathbb{R}^2$ having sides parallel to one of the sides of the polygon. Then:

Theorem 2. There exists a constant $C$, independent of $N$, such that

$$\mu\{x: Mf(x)>\alpha\} \leq C(\log N)\frac{\|f\|_2^2}{\alpha^2}.$$

These results are sharp, and they explain the different behavior of $S_0$ and $S_\lambda$, $\lambda > 0$ in Theorem A.

Theorem 1 suggests naturally the following question: Is there a polygonal region $D$, whose sides have infinitely many directions, such that the operator given by $\widehat{Tf}(\xi) = \chi_D(\xi)\hat{f}(\xi)$ is bounded on some $L^p$, $p \neq 2$?

Let $\theta_1 > \theta_2 > \ldots$ be a decreasing sequence of angles, with $0 < \theta_i < \frac{\pi}{2}$, and consider the region $P_\theta$ of the figure.

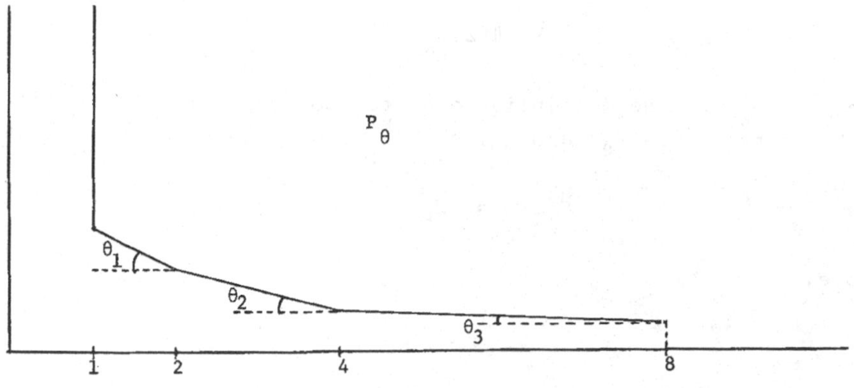

The multiplier

$$\widehat{T_\theta f}(\xi) = \chi_{P_\theta}(\xi)\hat{f}(\xi),$$

and the maximal function

$$M_\theta f(x) = \sup_{x \in R \in B_\theta} \frac{1}{|R|}\int_R |f(y)|dy,$$

where $B_\theta$ = {rectangles of arbitrary eccentricity oriented in one of the directions $\theta_i$}.

Claim. Boundedness properties of $M_\theta$ and $T_\theta$ are equivalent in a very precise sense.

(A). The boundedness of the operator $T_\theta$.

We shall show now under the assumption that $M_\theta$ is a bounded operator on $L^{(p/2)'}(\mathbb{R}^2)$, that $T_\theta$ is also a bounded operator on $L^q(\mathbb{R}^2)$, with $q \in (p',p)$.

One of the main ideas here is to invoke the inequality

$$\int |\tilde{f}(x)|^2 w(x)dx \le C_s \int |f(x)|^2 A_s w(x)dx, \qquad \text{(see [8])}$$

where $A_s w(x) = [(w^s)^*(x)]^{\frac{1}{s}}$, and $*$ denotes the Hardy-Littlewood maximal function.

Suppose that $E_k = \{(x,y) \in \mathbb{R}^2,\ \ 2^k \le x \le 2^{k+1}\}$, and let us consider the operators

$$\widehat{S_k f}(\xi) = \chi_{E_k}(\xi)\hat{f}(\xi).$$

Then we can use the Littlewood-Paley theory to obtain:

$$\|Tf\|_p \simeq \|(\Sigma|S_k Tf|^2)^{\frac{1}{2}}\|_p.$$

However, if $H_k$ is the multiplier operator corresponding to the half-plane $F_k$ tangent to $P_\theta$ along its $k^{th}$ side, we have

$$S_k Tf = H_k Tf.$$

Therefore,

$$\|\Sigma|H_kS_kf|^2\|_{\frac{p}{2}}^{\frac{p}{2}} = \sup_{\|w\|_{L^{(\frac{p}{2})'}}\le 1} \sum_k \int |H_kS_kf(x)|^2 w(x)dx \le$$

$$\le C \sup_w \sum_k \int |S_kf|^2 w^*(x)dx \le C_p\|f\|_p^p$$

where $w^* = \sup_k [m_k(w^{1+\varepsilon})]^{1/1+\varepsilon}$ and $w_k$ denotes the Hardy-Littlewood maximal function in the direction $\theta_k$.

Q.E.D.

(B) We shall now show that if $T_\theta$ is bounded on $L^p$, $p > 2$ then $M_\theta$ is of weak type $((p/2)', (p/2)')$, modulo some tauberian condition.

First we show under the assumption that $T_\theta$ is bounded on $L^p(\mathbb{R}^2)$, we have the following Meyer Lemma:

$$\|(\sum_k|H_kf|^2)^{\frac{1}{2}}\|_p \le C_p\|(\sum_k|f_k|^2)^{\frac{1}{2}}\|_p,$$

where $H_k$ represents the Hilbert transform in the direction $\theta_k$.

To see this, it is enough to work with finite collection of smooth functions, $f_1,\dots,f_N$ and we may also assume that each $\widehat{f_j}$ has compact support. Then we look for estimates with constants $C_p$, independent of these assumptions.

Let us expand $P_\theta$ by a convenient factor $\rho$ so that

$$\sup_j(\text{diam}(\text{Supp of } \widehat{f_j})) \le \rho/2.$$

Then, for each $j = 1,\dots,N$ there exists $w_j$ so that

$$\widehat{H_kf_j}(\xi) = \chi_{P_\theta^\rho}(\xi+w_j) \cdot \widehat{f_j}(\xi) = \chi_{P_\theta^\rho}(\xi+w_j)\left(\widehat{e^{-iw_j x} f_j}\right)(\xi+w_j)$$

$$= \widehat{e^{iw_j\cdot x} T_\theta^\rho(e^{-iw_j\cdot x} f_j)},$$

where $T_\theta^\rho$ is the multiplier associated with the characteristic function of $P_\theta^\rho$. Therefore,

$$\|(\sum_k|H_kf_k|^2)^{\frac{1}{2}}\|_p = \|(\sum_j|T_\theta^\rho(e^{-iw_j\cdot x} f_j\|^2)^{\frac{1}{2}}\|_p \le C_p\|(\sum_j|f_j|^2)^{\frac{1}{2}}\|_p.$$

Q.E.D.

Suppose now that $\{R_k\}$ is a collection of rectangles in $B_\theta$ having the following property:

$$(P)\quad \forall k,\ |R_k \cap \bigcup_{j<k} R_j| \leq \tfrac{1}{2}|R_k|.$$

Then the following estimate must be true

$$[vi]\quad \|\Sigma \chi_{R_k}\|_{\frac{p}{2}} \leq C_p |\cup R_k|^{\frac{2}{p}}.$$

<u>Proof</u>. Consider $E_k = R_k - \bigcup_{j<k} R_j$; then $|E_k| \geq \frac{1}{2}|R_k|$, by hypothesis (P), which implies that

$$|H_{i_k}\chi_{E_k}(x)| \geq \frac{1}{100} \text{ on } R_k^{(1)} \qquad \text{(see figure)},$$

where $H_{i_k}$ denotes the Hilbert transform in the direction of $R_k$.

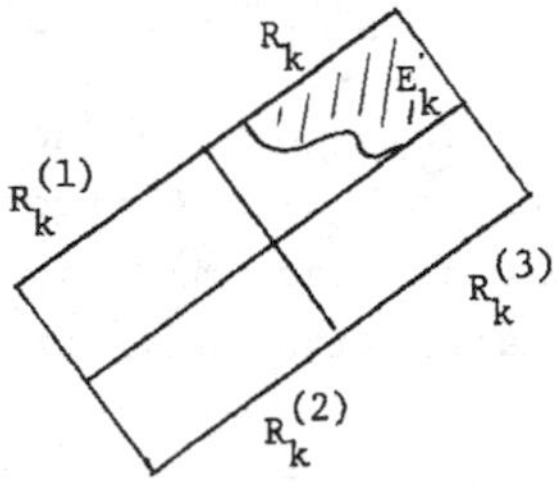


If we denote by $H_{i_k}^\perp$ the Hilbert transform in the perpendicular direction, we obtain,

$$\chi_{R_k}(x) \leq \left| H_{i_k}^\perp\left(\chi_{R_k^{(3)}} H_{i_k}\left(\chi_{R_k^{(2)}} H_{i_k}^\perp\left(\chi_{R_k^{(1)}} H_{i_k}\chi_{E_k}\right)\right)\right)(x)\right|.$$

Therefore we can invoke Meyer's Lemma to conclude the proof of the covering lemma [vi].

Suppose, for example, that we know that $M_\theta$ verifies the following estimate: $|\{M_\theta \chi_E(x) > \frac{1}{2}\}| \leq C|E|$ for every open set $E$. Then property [vi] will imply that $M_\theta$ is of weak type $(p/2)'$ automatically.

It is now an interesting question to decide for which families of $\{\theta_k\}$ the operators $M_\theta$ or $T_\theta$ are bounded on some range of $L^P$ spaces. Is there any geometric characterization of good sets $\{\theta_k\}$ of directions? Basically our present knowledge is contained in the following examples: (1°) If the sequence $\{\theta_k\}$ is lacunary then $T_\theta$ is bounded on every $L^p$ space, with $1 < p < \infty$, and $M_\theta$ on those with $1 < p$. (2°) If $\theta_k \simeq k^{-n}$, $(n = 1,2,\ldots)$ then $T_\theta$ is bounded only on $L^2(\mathbb{R}^2)$ and $M_\theta$ is bounded only on $L^\infty(\mathbb{R}^2)$. Here, the enemy is the Kakeya set. (See [1], [2], [4], and [7].)

[C] Further remarks. Recently, A. Ruiz [5] has obtained versions of Theorems (A) and (C) for more general types of curves in $\mathbb{R}^n$, enabling him to give a negative answer to a problem of N. Rivière: Is the fundamental solution

$$m(x,y) = \frac{1}{x^2-y+i}$$

of the Schrödinger operator a Fourier multiplier of $L^p(\mathbb{R}^2)$, $p \neq 2$? This result has been obtained independently by C. Kenig and P. Tomas by a slightly different method [3].

## References


[1] Córdoba, A. and Fefferman, R., On the equivalence between the boundedness of certain classes of maximal and multiplier operators in Fourier analysis, P.N.A.S., USA, 1977.

[2] ———, On differentiation of integrals, P.N.A.S., USA, 1977.

[3] Kenig, C. and Tomas, P., to appear.

[4] Nagel, A., Stein, E., and Wainger, S., Differentiation along lacunary directions, P.N.A.S., USA, 1978.

[5] Ruiz, A., Thesis, University of Madrid.

[6] Córdoba, A., The multiplier problem for the polygon, Annals of Math. 1977.

[7] Stromberg, J., Maximal functions for rectangles with given directions, Mittag-Leffler Inst. (1976).

[8] Córdoba, A. and Fefferman, C., A weighted norm inequality for singular integrals, Studia Math. 57 (1976).